the kitchen as laboratory

arts and traditions of the table: perspectives on culinary history

the kitchen

reflections on the science
of food and cooking

Edited by
César Vega, Job Ubbink,
and Erik van der Linden

Foreword by Jeffrey Steingarten

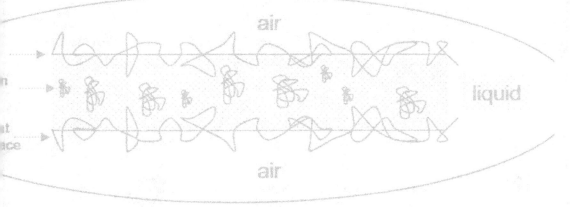

as laboratory

COLUMBIA UNIVERSITY PRESS   NEW YORK

Columbia University Press
Publishers Since 1893
New York   Chichester, West Sussex
cup.columbia.edu
Copyright © 2012 Columbia University Press
Paperback edition, 2013
All rights reserved

Library of Congress Cataloging-in-Publication Data

The kitchen as laboratory : reflections on the science of food and cooking / edited
by César Vega, Job Ubbink, and Erik van der Linden.
     p. cm. — (Arts and traditions of the table)
  Includes bibliographical references and index.
   ISBN 978-0-231-15344-7 (cloth)—ISBN 978-0-231-15345-4 (pbk.)—
ISBN 978-0-231-52692-0 (e-book)
   1. Food—Analysis.   2. Food—Composition.   3. Cooking.   I. Vega, César.
II. Ubbink, Job.   III. Linden, Erik van der.
  TX541.K55 2012
  664'.07—dc23

                                      2011029237

Book and cover design by Milenda Nan Ok Lee
Cover image by StockFood / Ralf Mueller

The use of recipes, preparation of food, testing of scientific concepts, and application of techniques described in this book must be done by only qualified
people, and all possible safety precautions should be followed.

CONTENTS

FOREWORD

JEFFREY STEINGARTEN

Are we in the midst of a culinary revolution? Will cooking ever be the same? The answers are, respectively, yes and no. But what kind of revolution is this anyway? We cannot even agree on its name. We do not know who started it or when. And we do not know where it is heading or what the world will be like when it has run its course.

The previous revolution was launched in 1972 by Henri Gault, a French journalist and critic. In an article, he called it "nouvelle cuisine" and the name stuck, although by today's sophisticated branding standards, it was too general, ambiguous, unspecific, and bland. The term *nouvelle cuisine*, however, took off like a rocket and, as far as I know, was never seriously challenged except by people who challenge everything and those who considered the cuisine unrevolutionary and not nouvelle enough. French chefs at least as far back as François Pierre La Varenne in the seventeenth century have claimed to be revolutionizing French cooking through a process of simplification and purification. In 1973, in "Vive la nouvelle cuisine française," Gault and fellow critic Christian Millau issued the Ten Commandments of nouvelle cuisine, of which only some specified a new way to think about food; one commandment, for example, recommended introducing air-conditioning into restaurant kitchens. But when a journalist aims for a catchy list of ten or twenty items, he or she must often pad the list to reach the magic number.

Gault and Millau were, in essence, reporting on how prominent young chefs were actually cooking in a new way. But a renewal of sorts had been bubbling through the old style of French haute cuisine in, for example, the cooking of the great Fernand Point. At La Pyramid, his restaurant in Valence, which he opened shortly after World War I, he trained chefs later associated with the "nouvelle cuisine," particularly Paul Bocuse, Michel Guérard, and Alain Chapel. What made the young generation's approach to cooking and food presentation a revolution was simply that a journalist called it a revolution and gave it a catchy title.

The problem is that the name of the current gastronomic revolution was coined by scientists instead of journalists, a mortal error. Nonetheless, the term they first produced, *molecular gastronomy*, was perfectly fine. For an appellation to be perfect, I suppose, it must include everything it names and exclude everything it does not. But names are rarely perfect. By this standard, the term *molecular gastronomy* is no worse than, say, the titles *As You Like It* and *All's Well That Ends Well*, which indicate nothing about the story lines of the plays. "Molecular gastronomy" is a metaphor or, really, a metonymy. When my wife, Caron, was learning Mandarin in graduate school, she informed me that the modern Chinese word for a ballpoint pen is *yuan zi bi*, which means "atomic pen." Now, do the Chinese people believe that there is a nuclear bomb inside each pen or that the ballpoint pen represents a novel exploitation of the atom? Of course not. For them, at that remote time when the ballpoint pen thrilled them with its novelty, *atomic* seemed the perfect word to express the idea of futuristic. Maybe we should call the current revolution atomic gastronomy.

In this handsome book, the editors suggest that we refer to this movement as *science-based cooking*. This term, however, reveals nothing about what went on in restaurants such as Ferran Adrià's El Bulli, where I had my first meal in 1998. What impressed me most about this meal was a plate arrangement that had the main ingredient in the center, enclosed in a circle of other ingredients, one of which was the inside of a tomato—the gel and seeds that French cookbooks have you squeeze out and discard from the start, but here presented intact. A day or two later, I asked Ferran how he had accomplished this, and he showed me, holding a long plum tomato in one hand and a knife in the other. After slicing off the top of the tomato, he ran the knife down the outside of the flesh (pericarp) wherever it was supported by one of four internal ribs (septa), cutting through each rib. The result was six or eight sections of tomato flesh that he pulled away, revealing four compartments (locular cavities) packed with gel and seeds that he removed, intact, with a spoon. Nobody else of whom I am aware has ever handled a tomato like this. (If you follow Adrià's method to separate the inside from the outside of a raw tomato and taste both, you will learn that the stunning umami flavor of the tomato lies mainly in the seeds and gel, as was later demonstrated by Heston Blumenthal.) Things change when you cook the tomato. Another example is Ferran's cauliflower couscous, which is created with a sharp, thin knife in one hand and a cauliflower floret in the other, held over a sheet of waxed paper. When you "peel" off the outer ⅛ inch or slightly more of the cauliflower, it falls onto the paper in tiny white balls that resemble couscous. Great amusement ensues as the diners discover the delicious trick that has been played on them.

There is nothing strictly scientific about it. Harold McGee prefers the phrase *experimental cuisine*. I like that term, too, because it is inclusive, and I believe that it implies a systematic series of hypotheses and trials. The techniques for separating the tomato and shaving the cauliflower were, as far as I can tell, unprecedented—but not quintessentially scientific. Both depend on the close observation of natural forms, the province of either biologists or painters and engravers.

When did it all begin? A crucial moment came in 1974 when French chef Pierre Troisgros asked fellow chef Georges Pralus to figure out how he could cook his foie gras terrine without losing the typical 30 to 50 percent of its weight. Pralus turned the industrial method of *sous vide* cooking to gastronomic ends, wrapping the terrine in plastic and immersing it in hot water that was maintained at a moderate temperature. And when it was done, the terrine had lost only 5 percent of its weight. History was made! Profits soared! Nearly all of the foie gras's fat was kept in the terrine, still integrated with the liver, so the calorie count in each serving size must have soared. I wonder if anybody bothered to notice this.

Deploying large-scale industrial technology to further small-scale gastronomic goals has been seen again and again in the development of the new gastronomy. (Let's call it that for now.)[1] The

1 The new gastronomy is a flawed designation. According to the *Oxford English Dictionary*, gastronomy is the "art and science of delicate eating," but we are concerned with the making of food, with cooking, with cuisine, and with eating. Perhaps we should refer to the movement as the new cuisine. But I believe that that name has been taken.

alginates, employed with spectacular success in spherification, have long been used in the faux-food industry. When a puree of sour cherries (which can economically include scraps and shards and squashed fruit) is gelled, it can be turned into cherry-like objects for filling pie crusts; strips of pimento are produced in a similar manner for stuffing olives. Centrifuges and rotary distillers can extract the very essence of ingredients. Freeze drying produces useful results. The irony of importing the instruments and tools of the processed-food industry and its "additives" into the artisan's kitchen has been lost on nobody.

Which bring us to the volume in your hands, *The Kitchen as Laboratory*. It is in many senses a sequel to three works. The first is a lecture delivered in 1969 by Nicholas Kurti, a professor of physics at Oxford University, to the august fellows of the Royal Society. In "The Physicist in the Kitchen," Kurti demonstrated some ways in which the use of scientific knowledge and techniques can improve the state of cooking. This topic was without precedent, but Kurti's towering achievements in low-temperature physics (his was the first laboratory to attain a temperature of only one-millionth of a degree above absolute zero) undoubtedly justified to his audience his interest in cooking and the science underlying it.

In 1984, Harold McGee's *On Food and Cooking* was published. It was a detailed survey of what was then known about the science of food and cooking, and with it, McGee shifted the culinary zeitgeist, the "thought-spirit" of the field. It was what we now refer to as a game changer. Within a decade of the book's publication, nearly any English-language food writer who understood nothing about the science underlying his or her subject was beginning to be viewed as a bit out of step and his or her writing as incomplete.

And then in 1988, Kurti and his wife, Gianna, published *But the Crackling Is Superb*, a collection of essays similar to this volume except that the scientist-authors were not experts in food science or, for that matter, in gastronomy. The Kurtis had solicited manuscripts from all members of the Royal Society, among whose fellows are counted the most accomplished scientists in England. Among the contributions of fellows no longer living, the essay by Benjamin Thompson, Count Rumford, on brewing perfect coffee would have been brilliant—if it had worked. Without wanting to denigrate the Kurtis' pioneering project in any way, some other fellows, both living and dead, also reached disappointing conclusions.

The Kitchen as Laboratory is a banquet of a different sort. The editors—César Vega, Job Ubbink, and Erik van der Linden—have asked scientists already engaged in the study of food—already fascinated by food, even addicted to food—to contribute essays on their particular areas of inquiry, each with practical implications for life in the kitchen, particularly the home kitchen. I have noticed that the main way in which I form a first impression of a new cookbook (or sometimes an old one) is whether, when I leaf through it, I am overcome by an irresistible desire to rush into the kitchen and start to cook. And while I am always delighted by the lucidity of fine scientific writing and find the analysis of culinary phenomena to be captivating in itself, my favorite contributions to this collection are those that have sent me into the kitchen.

Grilled cheese, as offered in chapter 1, is irresistible, as is cheese fondue, although in recent years the popularity of fondue has dimmed because most versions seem heavy and because it often separates and clots. Both Swiss chefs from the haute cuisine whom I profiled in my article about cheese fondue in *Vogue*, and who make fondue only for themselves and only occasionally, use only Vacherin fribourgeois (a creamy washed-rind cheese), no kirsch, and not much heat; but even following their advice, the outcome was not perfect. Chapter 1 has given me the strength to try again.

The history in chapter 4 of spherification, as it was developed at El Bulli by exchanges between cooks and scientists, shows that whatever becomes of the recipes of Ferran Adrià, there is no doubt that spherification will hold its own next to more conventional techniques in the kitchen. This I also expect for several other of the novel techniques, such as *sous vide* cooking.

I was fascinated to learn from our expert in foam, in chapter 14, that the reasons why egg whites beaten in a copper bowl form a larger and more stable foam have not yet been fully understood. Did you know that this is still an open question? What will it take for the contributors to this collection to figure it all out?

In chapter 23 (in an homage to Kurti), the dilemma of turning duck skin crisp without incinerating the interior flesh may have been solved. But I am just as interested in applying the findings to geese. A crackling goose leg and its thick crunchy skin put every crispy duck to shame.

And chapter 25, which demonstrates some potential advantages for extracting the fat-soluble flavors of coffee, opens up an intriguing realm of possibilities that Count Rumford never imagined. Now I must leave this discussion and get into my kitchen. There is much work to be done!

And where is it all heading? At last January's meeting of Madrid Fusión, where in the past, new cooking techniques, especially those developed with the help of scientists, have been introduced to the general public, two novel examples that were presented combined very low temperatures and very high pressure. The equipment required—not brought into the lecture hall but shown in photographs—was the size of a school bus. This said to me that in the near future, no huge leaps in science-based cuisine—comparable to the popularity of liquid nitrogen and hydrocolloids—can be foreseen. But existing techniques have a long way to go before they will be exhausted. I guess that this sounds like a quote attributed to a commissioner of the Patent Office, who announced, in the mid-nineteenth century, that everything useful had already been invented.

ACKNOWLEDGMENTS

This book, envisioned in 2007 and begun in earnest in late 2008, represents a journey that has fostered our professional and personal growth, while strengthening our three friendships—even though sometimes there were too many cooks in the kitchen!

It has been a long journey.

During the various editing phases of the book, we felt challenged and energized by the feedback, help, and advice of all our contributors. We want to express our admiration and gratitude to all of them. The unique nature of the book allowed them to create their own pieces, reflecting their personal approaches. At the same time, all our contributors were highly understanding of the occasional hitch in our editing of the book. This book is as much theirs as it is ours!

The role our anonymous reviewers played in shaping the final version of this collection was indispensable. Their thoughts and suggestions are the salt and pepper of the book. Without them, this anthology would have been insipid, one-dimensional; it would have been missing that secret ingredient. Thank you for you candor and for believing we were on to something.

Managing more than thirty chapters with more than fifty contributors dulled our ability to ensure the chapters flowed in a harmonious, understandable way. This was where very talented people—our friends—came to the rescue. Not only did they improve the quality of the manuscripts they proofread, but they did it under very tight schedules—and for this we thank Anne E. McBride, Patricia Gadsby, Catherine Kwik-Uribe, Alice Lee, Mina Roussenova, Adam Burbidge, Hugo Perez, Amanda Kinchla, Jan Engmann, and Sam Townrow.

On more a personal level, each of us offers a few words:

It has been close to five years since I seriously embarked on my personal journey in the field of the science of cooking. A few people have been immensely important in setting the stage for me to make a valuable contribution to this area of science. I am grateful to Christopher Young. Since he invited me for a short, yet significant, "internship" at the Fat Duck in Berkshire, England, and through to his very recent modernist-cuisine odyssey, we have engaged in many mind-twisting discussions around what food could be. I thank Michael Laiskonis— chef, artist, friend. His sensitivity and equanimity taught me that food is as much art as it is science. I thank Anne E. McBride, a never-stopping mind and a true gourmet, with whom I have enjoyed some of the most rewarding culinary experiences of my life. I am extremely grateful for her advice when I needed it the most and for her always refreshing bluntness.

Finally, and most important, I thank my wife, Elizabeth, for her infinite patience before, during, and after the making of this book. She knows the ins and outs of it without even reading a single paragraph, which reminds me how boring I can be at times. She knows that cooking and science are more than a passion for me, and she is my biggest fan. The day this book rests on our bookshelf, please remember that it is for the most part, because of you. Let's make this one more bedtime reading for Lucas!

CÉSAR

Even though now more than twenty years have passed, I still feel it is appropriate to first thank my former colleagues at the restaurant De Beukenhof, where as a young student, I worked for several years and was first exposed to gastronomic cooking. The passion and craftsmanship of the kitchen and restaurant brigade sparked in me a deep and continuing interest in food and cooking . . . and of course, at the time, I had the best moonlighting job imaginable!

During the past couple of years, I have not only learned a lot from my friends in the experimental kitchen at the Nestlé Research Center in Lausanne, but I have also spent many enjoyable evenings in our common pursuit of innovative cooking. Fernand Beaud, Laurence Donato, Olivier Roger, and Delphine Curti—thank you for your friendship, the inspiring interaction, and the many culinary ideas we have explored together.

Finally, Marijke, without you my life, but definitely my food and cooking, would be much less inspiring. I tremendously appreciate that you allow me to indulge in shopping sprees at the street markets here in Lausanne and Basel, in neighboring France, and in fact wherever there is an opportunity. However, you also fully participate in trying to find the best seasonal and local foods available. I greatly enjoy our cooking together, and if there is one thing that you have taught me, it is that it can be worthwhile to occasionally open a cookbook!

JOB

I would like to thank many different people, here. I thank Hervé This for introducing me to the field, for his support, and for our discussions. I thank Jan Groenewold and Eke Mariën for organizing our Science Is Cooking events at Wageningen University in the Netherlands. I thank Ralf Hartemink for his support of the master of science courses on molecular gastronomy. I am very grateful to Fons Voragen for working out the first course with me and to Gerrit Smit and Peter Wierenga for their involvement in the courses as well. I thank René Hoogerwerf, Rudolf Barkhuizen, and Ingmar van Borselen for their endless enthusiasm in taking care of the practical aspects of the course work, along with Jean Paul Vincken, Silvia van Kempen, Martine van Gool, Elke Scholten, and Guido Sala. I further acknowledge the continuous support from and discussions and/or collaborations with Kees de Gooijer, Kees de Graaf, Heston Blumenthal, Jonnie Boer, Remko Boom, Aalt Dijkhuizen, Allen Foegeding, Rolf Hilfiker, Helen Hofstede, Els Jansen, Julian McClements, Harold McGee, Sanne Minten, Sylvia Posthumus, Moshik Roth, Jorge Ruiz, Sidney Schutte, Robyn Stewart, Paul Venema, Thomas Vilgis, David Weitz, Pierre Wind, and Heleen and Jurgen Zuiderduin.

Finally, I want to attempt to express my gratitude to my wife, Alexandra, and our children, Isabelle and Lucas, and to convey how important and inspirational they are to me. Alexandra, I thank you for your continuous support in many ways (including, of course, working on this book during weekends), your enthusiasm, ideas, and engagement in all our discussions on the diverse matters of science and other topics, and for the many things you taught me about chemistry and food. You inspire me a great deal. Isabelle and Lucas, of course you know by now that science can help us understand what happens to food and how we perceive it. Thank you for inspiring me with your questions, for sharing your ideas, and for your conversations. You make me a proud father. Please keep asking "how?" Alexandra, Isabelle, and Lucas, I hope that reading this book will give you enjoyment and inspiring insights for the future.

ERIK

the kitchen as laboratory

INTRODUCTION

The Case for Science Inspired by the Kitchen

CÉSAR VEGA, JOB UBBINK, AND ERIK VAN DER LINDEN

this book is a culinary anthology, a dream come true. In it is a collection of carefully selected stories that relate to food, its preparation, its perception—and how we eat it. However, what distinguishes this collection from others is the fact that it is infused with science. Our objective is to help the reader better understand how food is transformed during cooking and eating. The essays are as diverse as the foods they aim to describe: from simple foods, such as chocolate chip cookies, grilled cheese sandwiches, pizzas, and soft-boiled eggs, to

> Gastronomical knowledge is necessary to all men, for it tends to augment the sum of happiness.
>
> Jean Anthelme Brillat-Savarin,
> *Physiologie du goût*

foods of a higher level of complexity (both culinary and scientific), such as sauces, sugar glasses, and jellified beads. Some of the stories even touch on food and culture, and on the relationship between food and society. The discussion incorporates how personal background, one's culture, dining companions, eating environment—the lighting, the occasion, and the like—affect the way foods are perceived. Similarly, some of the essays address cooking methods: from the characteristic high-temperature wok-sautéing of Chinese cooking to slow cooking in temperature-controlled water baths and distillation-based aroma extraction and concentration. To some readers, the diversity of the topics and approaches of the stories might take away from the cohesiveness expected in a book. To us, as editors, the beauty of the book is in the common thread that runs throughout: the quest to transform personal culinary observations and reflections into scientific knowledge.

Why did we choose to compile a book like this? Why now?

Over the past few years, an interest in understanding the chemical and physical phenomenon of cooking—what some call the science of cooking—has rapidly increased in popular and technical literature. The three of us, working as food scientists, were already personally and professionally involved in the field and decided to pursue a more serious, disciplined, and at the same time, entertaining exploration of the interface between science and cooking. We soon realized that this needed to be a collaborative effort, and we are delighted to have contributions from scientists, chefs, and food lovers from around the world.

Our interest in the field of science-based cooking arises from many different influences. On the one hand, we find it fascinating that by using scientific concepts, we can better control classical cooking while further exploring the creative opportunities that existing food ingredients and preparation techniques afford. On the other hand, over the past decade or so, many new techniques and ingredients have been introduced into the professional kitchen, dramatically transforming the landscape of restaurant cooking. For us, as food scientists, it is of interest to note that many of these techniques and ingredients were originally developed for industrial-food manufacturing. Industrial techniques, such as freeze drying, centrifugation, inductive heating, and vacuum packaging, and "new" ingredients, such as native and modified starches, alginates, xanthan and gellan gums, carrageenans, and "meat glues," are now frequently applied in the restaurant kitchen. Their use, however, is often with only a superficial understanding of how they work. Despite this lack of understanding, chefs have prepared dishes that combine new or modified techniques with new ingredients that deliver previously unknown eating experiences—not always to everyone's liking but interesting nonetheless!

Take, for example, the black sesame sponge cake served in 2008 at El Bulli, the world-renowned restaurant located in Roses, Spain. The sponge batter is primarily composed of eggs, black sesame seed paste, sugar, and just a bit of flour. Once all the ingredients are thoroughly combined, they are poured into a whipped-cream dispenser and charged with nitrous oxide, a gas necessary for the aeration and subsequent expansion of the mixture during cooking. A 6-ounce (180 mL) plastic cup, filled one-third of the way, is placed in the microwave for about 45 seconds. The result is the lightest, fluffiest sponge cake you can imagine! The final volume is larger than the volume of the cup; that is, it expands by more than 200 percent. This amazing amuse-bouche (literally, "mouth amuser": a complimentary tidbit offered before the main course) is made possible by the unconventional use of technology—efficient aeration via the dispenser and quick cooking with microwaves—and an understanding of the basic role of each ingredient in the recipe. Too much flour makes the batter too heavy for expansion; hence the final structure, the scaffolding that holds the foam, is built through the coagulation of eggs.

However, to take the best creative advantage of these new technologies and ingredients, it is imperative that the cook understands what happens to food during preparation and how this may be influenced not only by the ingredients but by the cooking methods, conditions, and compositions. Therefore, many cooks, food professionals, and amateurs alike are increasingly founding their cooking on a systematic and scientific understanding. Such understanding is generated by testing hypotheses on the behavior of foods. We cannot truly control what we do not understand. Science helps us gain this control with answers to questions such as: Why is it not possible to whip cream when it is warm? What is whipping cream to begin with? How does cream of tartar help stabilize a meringue? Why is it that you can make tequila sorbet only with liquid nitrogen and not in a conventional ice cream maker? Even better, how can you make ice cream in 30 minutes without the help of a freezer and a mixer? It follows that from being just a place to prepare food, the kitchen is evolving into a place where cooks increase their understanding of food—by means of observation, measurement, and record keeping—in other words, a laboratory.

Further motivation behind the development of the science of cooking can be traced back to our own kitchens, where more than once, we have suffered from the disappointment of a failed recipe. "It worked last time!" we tell ourselves. "Why, then, didn't it this time?" More often than not, the answer can be found in the lack of quantitative precision reigning in the kitchen. This makes cooking, particularly baking, a fragile process in which inconsistency and failure are more the norm than the exception.

Why did we choose to publish this book as an anthology? Inspiration came in the form of a book published more than twenty years ago: *But the Crackling Is Superb: An Anthology on Food and Drink by Fellows and Foreign Members of the Royal Society*, edited by Nicholas Kurti and Giana Kurti. This book, a wonderful compilation of stories on the scientific aspects of food and cooking, was a call to action to scientists who, according to the Kurtis, tended to shy away from a serious application of their scientific skills to the understanding of cooking. In that regard, we believe that the Kurtis were ahead of their time—cooking was not yet seen as a topic worthy of serious scientific investigation. But the world has changed. The time is right for a collaborative book put together by scientists who are fervent cooks and by cooks who are curious scientists. This book is full of personal accounts that contribute to the understanding and practice of kitchen science; it is an homage to *But the Crackling Is Superb*.

A book format is of particular merit as the field is becoming saturated with blogs, magazine articles, and newspaper columns that attract a wide audience but are scattered, short-lived, and rarely discuss the real science behind a dish or cooking technique. A book such as ours represents a more lasting resource, providing a useful and interesting snapshot of the recent activities, ideas, and approaches in the field of science-based cooking.

We do not aim to fully cover the vast and rapidly expanding world of science-based cooking. We could have chosen to write a systematic book about the physics and chemistry of food and cooking or about the science of food perception. However, many other books, such as the excellent monographs by Peter Barham (2001), Harold McGee (1992, 2004, 2010), Hervé This (1993, 2006), and Thomas Vilgis (2005), already cover these subjects in one way or another. Our book focuses on a very limited selection out of a virtually infinite range of possible topics about the science of food and cooking. What makes it unique and interesting is not only the diversity but the meat of the contributions. The essays come from the minds of those who are culinarily curious, systematic, and, at times, philosophical. These stories are all personal accounts and thus convey a range of individual points of view. The book, with its many contributors, is therefore rich for its different ways of looking at a single subject.

The essays display a certain sequence with an underlying logic. This initially posed a challenge to us: how to structure an anthology that is diverse and yet incomplete. We spent three days in Leiden, the

Netherlands, in March 2010 to discuss, argue, and debate at length the best organization for such a sui generis collection. Food, needless to say, was an integral part of this meeting: we treated ourselves to an innovative and delicious dinner at chef Moshik Roth's Michelin-starred 't Brouwerskolkje. This restaurant turned out to be a particularly appropriate setting to conclude an important editing phase of the book because the chef makes judicious use of advanced tools, such as freeze dryers, to realize his vision of innovative and outstanding food.

Moreover, this memorable dinner made us realize that apart from conveying a passion for food and providing instruction on the science behind cooking, we wanted the essays to share another common aspect: complexity. *Food is complex.* This complexity may manifest itself across various dimensions: the intricacies of the chemistry and physics underlying the cooking process; the individual sensory expressions of flavor, aroma, texture, and appearance of a dish, which also depend on the number and quality of ingredients; the cooking technique and its execution; and the eating environment and situation (lighting, room temperature, time, dining companions, state of mind). One-dimensional eating experiences are generally not memorable. This is implicit in the very title of *But the Crackling Is Superb*: a dish, such as roasted chicken, that displays crackling and crisp characteristics at its surface while having a tender core is more complex and interesting to the tongue than a food, such as a biscuit or yogurt, that exhibits a homogeneous texture. In order to obtain such complex outcomes (crisp skin and succulent meat), the cook must have an in-depth understanding of the cooking process and how the food reacts to it. Much more is known about crispy *and* succulent roasts since the publication of *But the Crackling Is Superb* (for a taste, flip to chapter 23).

The essays in this book are organized in increasing order of complexity with regard to recipe, technique, or approach to the culinary problem. The detailed ordering of the chapters, however, is to some extent arbitrary.

We are convinced that the progress of science and cooking go hand in hand. Science can be illustrated by cooking experiments and cooking can be better understood through science. We hope this collection lends substance to future discussions you will have not only around the table but in front of the stove. For us, it has been an immensely rewarding experience, and we hope our book convinces you of the

relevance and pleasures of science-based cooking. So, put on your apron, unpack your tools, and fire up your imagination because . . .
. . . something is cooking . . .
. . . in the laboratory.

Further Reading

Barham, Peter. 2001. *The Science of Cooking*. Berlin: Springer.
Brillat-Savarin, Jean Anthelme. (1825) 1982. *Physiologie du goût*. Paris: Éditions Flammarion.
——. 1971. *The Physiology of Taste, or, Meditations on Transcendental Gastronomy*. Translated by M. F. K. Fisher. New York: Knopf.
Kurti, Nicholas, and Giana Kurti, eds. 1988. *But the Crackling Is Superb: An Anthology on Food and Drink by Fellows and Foreign Members of the Royal Society*. Bristol: Institute of Physics.
McGee, Harold. 1992. *The Curious Cook: More Kitchen Science and Lore*. New York: Hungry Minds.
——. 2004. *On Food and Cooking: The Science and Lore of the Kitchen*. 2nd ed. San Francisco: Scribner.
——. 2010. *Keys to Good Cooking: A Guide to Making the Best Foods and Recipes*. New York: Penguin Press.
This, Hervé. 1993. *Les secrets de la casserole*. Paris: Belin.
——. 2006. *Molecular Gastronomy: Exploring the Science of Flavor*. Translated by Malcolm DeBevoise. New York: Columbia University Press.
Vilgis, Thomas. 2005. *Die Molekül-Küche: Physik und Chemie des feinen Geschmacks*. Stuttgart: Hirzel.

one

THE SCIENCE OF A GRILLED CHEESE SANDWICH

JENNIFER KIMMEL

why do certain varieties of cheese make great grilled cheese sandwiches? The secret lies in understanding how the molecules within cheese influence the ooey-gooey melted goodness that is the essence of a perfect grilled cheese sandwich.

It all begins with the cow (or goat or sheep). After all, cheese, no matter the variety, gets its start from milk. Even though milk is made of 80 to 90 percent water (in most hoofed species), it is still a very good source of proteins (casein and whey), carbohydrates (lactose, or milk sugar), and minerals (especially calcium). These three components, along with milk fat, are the essential ingredients for making cheese. Proteins (primarily caseins) give cheese its structure and allow the fat and a small amount of moisture to be retained while the majority of the water is removed. Lactose provides a food source for the growth of bacteria, which lend individual cheese varieties their distinctive flavor. The calcium in the milk determines how the proteins interact

Little Miss Millet stood with her skillet
Some butter and two slices of rye
Which selection of cheese
Would make her grilled sandwich please
"The secret is within the science," she sighed.*

* Poetry based on the nursery rhyme "Little Miss Muffet," in Iona Opie and Peter Opie, eds., *The Oxford Dictionary of Nursery Rhymes*, 2nd ed. (Oxford: Oxford University Press, 1951), 323–324.

and this interaction ultimately dictates the softening, melting, and stretching characteristics of the heated cheese in a grilled sandwich.

Before milk is converted into cheese, the casein proteins are arranged in individual clusters, called micelles, which are suspended in what is known as the aqueous phase. They contain two-thirds of the milk's total calcium and have a net negative charge that prevents them from aggregating together. To convert milk into cheese, however, the proteins must aggregate and form a curd, trapping both fat and water. To achieve aggregation, the negative charge must be eliminated from the casein micelles. This is accomplished either by adding acid and neutralizing the negative charge or by adding an enzyme and cleaving the portion of the cluster that contains the negative charge. While the transformation from protein aggregation to cheese is complicated, the main steps include cooking the curd and draining the whey, followed by salting and pressing the curds together. Aging is the final step, which allows for structure and flavor formation.

The ideal cheese characteristic needed to make a grilled cheese sandwich is melt. Who does not love to cut into a hot grilled cheese sandwich and see smooth, creamy melted cheese oozing from between the slices of grilled bread. But why do some cheeses melt better than others? Why do certain varieties melt as homogeneous molten masses, while others as oily lumps? Again, we go back to the molecular interactions within the cheese, primarily the interactions between the casein proteins and the calcium. The casein proteins are held together in the micelles by calcium bridges, and the number of these bridges is influenced by the acidity of the cheese. As cheese ages, more of the lactose is converted to lactic acid, causing the pH of the cheese to decrease and become more acidic. This, in turn, causes a dwindling in the number of calcium bridges within the casein micelles as the calcium solubilizes and moves from its position among the proteins to the entrapped water within the curd. The fewer the number of calcium bridges, the greater the mobility of the proteins as their connections give way.

The loss of the calcium bridges allows for the casein proteins to become more soluble (that is, less held together in the micelle), which also helps to better bind the fat originally trapped in the cheese. Therefore, when the cheese is heated, the protein molecules are able to flow, resulting in a nice even melt. Moreover, the soluble protein molecules are able to interact with the oil droplets, preventing them from leaking out of the cheese and causing an oil slick to form on top. However, if the pH continues to decrease, then too much of the calcium is solubilized and the caseins collapse on one another as a function of their low-pH insolubility. This results in a curdy melt with free oil leaking from the cheese. If the pH is too high, then not enough calcium is solubilized and the

Finished with her skillet
Proclaimed Little Miss Millet,
"My masterpiece was certainly not a glitch.
The science of cheese
Provides the interactions that I need
To create the perfect grilled cheese sandwich."

casein proteins are held too tightly together by the greater calcium bridging, resulting in cheese that does not melt or flow. An appropriate pH for the good melting of cheese is in the vicinity of 5.3 to 5.5. It of course depends on how the cheese is made, but in general, this pH range balances the different protein interactions and allows for a good melt.

Examples of cheeses with good melting properties include Gruyère, Manchego, and Gouda. These varieties balance the final cheese pH to achieve both soluble calcium and soluble protein, resulting in a cheese mass that melts and flows upon heating while keeping the fat trapped within the matrix. The pH can be both perfectly balanced and too acidic in a single cheese variety. Take mild versus aged cheddar. While the mild cheddar melts evenly and maintains the fat within the matrix, the aged cheddar, because of its lower, more acidic, pH, will melt into lumps, releasing free fat.

Manufacturers of processed cheese have developed a method to decrease calcium bridging. They use salts (specifically, citrate and phosphate salts) to bind the calcium from within the casein micelles. To make processed cheese, natural cheese is cooked with these salts, decreasing the calcium bridging from within the casein micelles as the overall pH is increased. The same thing happens to cheese fondue when wine is added. The tartaric acid in wine binds calcium from within, affecting the bridging process.

THREE-CHEESE GRILLED CHEESE SANDWICH

2 slices rustic white bread
Grated Gruyère cheese
Grated fontina cheese
Grated mozzarella cheese
Butter

Lightly butter the outsides of the bread slices. Top the bread with an equal mix of grated Gruyère, fontina, and mozzarella cheese. Close the sandwiches. Grill until the cheese is melted and the bread is toasted.

Adapted from www.foodandwine.com

The result is a highly homogeneous mixture of protein, water, and fat that when heated, produces a smooth, creamy layer of cheese sandwiched between two slices of golden toasted bread.

Further Reading

Lucey, J. A., M. E. Johnson, and D. S. Horne. 2003. "Invited Review: Perspectives on the Basis of the Rheology and Texture Properties of Cheese." *Journal of Dairy Science* 86:2725–2743.

two

SOUND APPEAL

MALCOLM POVEY

french fries, or chips in Britain, are an evocative and ubiquitous aspect of life in many parts of the world. The most delicious french fries combine a crisp exterior with a soft and light interior—texture in this case is a defining characteristic. A mealy pear or a soggy apple will disappoint: crispness is expected of an apple, whereas more of a crunch is expected from a pear. Certainly, a stick of celery is expected to be crunchy; wilting and soft celery will find few takers. Crisp lettuce is definitely preferable to the flaccid variety. Carrots, when cooked, are transformed from crisp sticks to soft, sweet objects, a change associated with the destruction of the cell wall and the release of its contents. Cooks find it challenging to achieve a crackling pork skin, which some people find delicious and others objectionable.

The totality of the sensory experience of eating is amazingly complex, mediated as it is through all the senses. One aspect is mouthfeel, or texture, which according to Alina Szczesniak (1988), is composed of cohesiveness, density, dryness, fracturability, graininess, gumminess, hardness, heaviness, moisture absorption, moisture release, mouth coating, roughness, slipperiness, smoothness, uniformity, viscosity, and wetness.

Apart from this extensive list of texture attributes, one other part of the total sensory experience is noisiness. That the sounds made by food should be as much a part of the experience of eating as color, taste, smell, and mouthfeel seems obvious, especially in the case of crispy–crunchy–crackly foods. Crispy, crunchy, and crackly sounds are sensorially distinct in relation to other sensory attributes. It is notoriously difficult to get people to agree on the differences among crispness, crunchiness, and crackliness. Therefore, here we use the descriptors interchangeably. This is not the case in every language. (For example, in Chinese, there are eleven different words to describe crispiness, and the distinct sound associated with each word is discernible to the Chinese ear.) It is only very recently that we have begun to understand how fundamental sound is to the culinary experience (figure 1).

What do we hear when we eat? Where do the sounds come from? We have known for some time that some foods break down in the mouth through a series of jumps in the force exerted by the teeth. It has been shown by Jianshe Chen, Cathrine Karlsson, and Malcolm Povey (2005) that each of these force jumps is followed by a burst or pulse of sound.

A visual display of the sound produced by crispy–crunchy–crackly food, in which the sound pressure is plotted against time (figure 2), shows a characteristically spiky appearance. In contrast, similar visual displays of the sound of people speaking or of music show a continuously varying, wavy appearance.

Would you think a food that makes no sound in the mouth was crispy–crunchy–crackly? Probably not. Sound is essential if we are to evaluate food as having this quality.

In the laboratory, there is a difference between or-

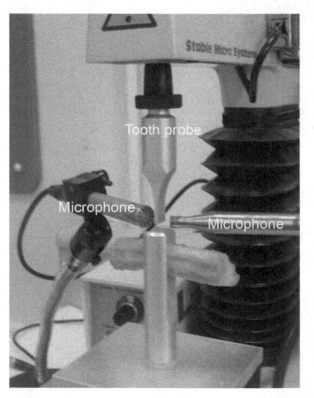

Figure 1 A french fry (chip) is tested for crispness by placing sensitive microphones close to the sample to capture the sounds it emits during fracturing.

dinary sounds of speech; music and ambient noise; and crispy, crunchy, and crackly sounds. Bite sharply into a very crisp apple and the resultant sound is immediately identifiable. The sound is so distinctive and evocative that the mouth of the hearer may start watering. It is not necessary to see or even be aware of the nature of the act of biting into an apple for this to occur. This was proven in a live interview

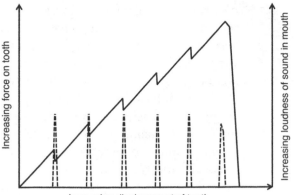

Figure 2 Ideal (smooth) and nonideal (fracturing) behavior of the type seen in crisp, crunchy, and crackly materials. The force on the biting tooth is plotted against the movement of the tooth through the food.

with a skeptical national radio host in Great Britain. In a studio in the north of England, I bit into an apple and then asked the host in the studio in London what had made the noise. "Apple" was the swift and laughing reply.

We should not have been surprised that, as a result of our experiments, we discovered crispy–crunchy–crackly to be a fundamental sensory descriptor, much like soft/hard, dark/light, sweet/umami (the pungent–savory taste associated with soy sauce and monosodium glutamate: it is the fifth basic flavor; the others are sweet, sour, bitter, and salty).

However, research has indicated some fundamental sound differences among crispness, crunchiness, and crackliness. Crunchiness tends to emit lower frequencies; in fact, the clue is in the sound of the word. Each is, in fact, onomatopoeic—imitative of the sound it denotes. So crisp (*crisssssp*) has high notes. Crackliness appears to combine features of both crispiness and crunchiness. Pork crackling will break in the mouth with a considerable *crunch*; however, as chewing proceeds, *crisp* notes are generated. This is because a well-cooked crackling will combine the hard, crunchy skin with the softer fatty subcutaneous layer, which forms a kind of foam, providing for many more fracture events during biting but requiring less force and creating a higher pulse rate.

When assessing the sensory characteristics of foods, taste panels may be trained to recognize, in a reproducible way, defined sensory characteristics. However, we used untrained taste panels and made no attempt to define crispness. We did this precisely because of the ambiguity of the language already described. Our taste panels were made up of university students who, by chance, came from a variety of backgrounds. Moreover, a majority spoke English as a second language. We asked the taste panels to rate eight different kinds of cookies (biscuits) on a scale from 0 (not crisp) to 10 (very crisp). We measured the sounds made by the volunteers as they bit into each cookie. To our amazement, we found a correlation of more than 90 percent between our volunteers' assessment of crispness and the rate and intensity of the sound pulses produced during biting.

We concluded that the universal definition of crispness is the rate and intensity of sound pulses produced in the mouth. That this is something innate was confirmed by Chen, Karlsson, and Povey when the results were reproduced with a trained taste panel. It has also been confirmed for raw and roasted almonds by Paula Varela and her colleagues (2007). Most food likes and dislikes are learned, contextualized, and therefore culturally influenced. However, crispy–crunchy sounds seem to have a significant innate component. Perhaps human evolution has equipped us with a preference for foods that make crisp noises because these sounds are generally indicative of fresher and more nutritious foods—roots, nuts, and green vegetables come to mind.

It seems crispy–crunchy sounds short-circuit ordinary hearing, perhaps being detected directly by the aural nerves. It may also be surmised that these sounds are interpreted differently in the brain. Interestingly, crispy–crunchy sounds can be extraordinarily intense. Our calibrated microphones indicated sound pressure levels far greater than what would be required to damage or even destroy human hearing. Yet we know that hearing is not damaged. Moreover, this provides strong evidence that crisp sounds are detected differently from ordinary sounds. Of course, we can modulate the frequency, pressure, and energy of the sounds by changing the rate at which we bite so that the sounds and the rate of biting interact through the brain. In addition, the aural nerves may not be as well coupled to the sound as the test microphones because of sound absorption in the mouth, thereby reducing the apparent sound pressure. Some children dislike foods that "pop" in the mouth because the sounds they make are too loud. One such food, a candy that appeared in the late 1970s as a follow-up to Pop Rocks, is Space

Dust. The powdered candy explodes in the mouth when it comes into contact with saliva.

Another surprise is the amount of ultrasound emitted by crisp foods as they are eaten. Ultrasound is sound frequency that exceeds the upper limit of human hearing, about 16 kHz. Experiments on deaf people have shown that although ultrasound cannot be heard, it can be felt, exciting areas of the brain adjacent to, but not coincident with, the hearing regions. Frequency analysis of the sound pulses generated by crisp foods shows the existence of frequencies over a very wide range that present roughly equally across the spectrum, from 2 Hz to 200 kHz. Mathematically, the presence of a very wide range of frequencies is necessary to have the simultaneous experience of a pulse and a characteristic sound. Otherwise, the frequency would manifest itself only as an ordinary sound whose intensity more or less gradually waxes and wanes. The presence and intensity of ultrasound was dramatically demonstrated in one of my lectures. I gave apples to members of the audience. They were asked to bite their apples at the same time. The collective bite was associated with an enormous burst of ultrasound, which was recorded by ultrasound-sensitive microphones. Projected onto a screen at the front of the lecture hall, the sound was given visual representation. Potato chips (crisps) produced an even bigger burst, to everyone's astonishment.

There is a big difference between sound transmitted through bone and through the air. In the air, sound is more rapidly attenuated and the highest frequencies are reduced the most, whereas in bone, the highest frequencies gain transmission. Therefore, most of the sound we "hear" when eating comes in the form of vibrations through the skull! It would be interesting to compare the eating experience of deaf people with that of hearing people. No doubt, deaf people can still "detect" the sounds in their mouths.

This is fascinating stuff and certainly makes us think about the enhancement of our enjoyment of food from the sounds they produce. Indeed, a heightened awareness of the different qualities of sound can greatly improve the experience of eating. Of course, many sounds other than crisp ones are important for the enjoyment of food. But outright noise, such as squeaky sounds—high pitched and sustained, like the scraping of a fork on a plate—are less desirable, of course; and the sound from chewing cheese curds is just . . . well, weird.

CRISP BATTERED FISH

Four 6-ounce (170 g) thick-cut haddock fillets
8 ounces (225 g) self-rising flour
10 ounces (300 mL) water
Vegetable oil, for frying
Seasoned flour (salt, pepper), for dusting the fish

Sift approximately 4 ounces (115 g) of the flour and a pinch of salt into a large bowl, then whisk in 7 ounces (210 mL) of the water until the batter is the thickness of heavy (double) cream. In a small bowl, sift the remaining flour and whisk in the remaining water to form a thick, almost pasty, liquid. Then stir this pasty liquid into the batter to create an uneven texture.

Heat the oil to 365°F (185°C), using a thermometer to ensure that this temperature is achieved precisely.

Dust the fish with seasoned flour and shake off the excess. Dip two fillets into the batter, ensuring that lumpy bits adhere, and fry for 8 minutes until golden and crisp. Repeat for the remaining fillets; make sure the oil remains at 365°F (185°C), and no higher, throughout.

Experiments with This Recipe

To test for crispness, place the battered fish on the texture analyzer platen and record the force displacement and acoustic output as the probe pierces the batter. The batter produced in this manner can be compared with a more traditional batter process in which 8 ounces (225 g) of plain flour is combined with 10 ounces (300 mL) of water. The self-rising flour introduces bubbles into the batter, which together with the uneven coating produced by the lumpy mixture, increases the number of interfaces in the batter that can break and therefore increases the "noisiness" and hence the pleasure of the eating experience.

The impact of sound is dependent not only on how we actually perceive the signal but on our expectations. This is also the case with perceptions like taste and color. We performed an experiment to test perceived taste and color by switching the color cues in candy. Green candy with strawberry flavor was often assessed as tasting of lime! Similarly, sounds produced by food packaging can predispose the consumer toward expectations of a particular texture. Food manufacturers consciously or unconsciously create packaging that reflects the noise their foods produce. So food that is liked for its crispiness is put into packaging that also makes crisp or crinkling sounds.

The potato chip is perhaps an example of how the more recent art of cooking (about 10,000 years old) fools our senses. Certainly, the early human diet was nothing like the one that has evolved over millennia and we enjoy today. But in another radio interview, this time on German radio, the researcher discovered that the more crisp noises a sausage made, the better quality it was!

It is time that in our sensory experience of food, we acknowledge the importance of sound, along with color, taste, smell, and texture. A heightened awareness of the sound foods make increases our eating enjoyment, enhancing life itself. Chefs, like Heston Blumenthal of the Fat Duck in Berkshire, England, have realized this, using techniques to amplify the texture and therefore the sound of their dishes (Blumenthal's techniques have been showcased on several BBC television cooking series). An apple must be judged in terms of how crisp it is, which is as important as its taste. The mange-tout, or snap pea, must also be crisp. The potato chip must be crunchy outside and light and soft inside. The traditional British meal of fish and chips combines crispness and crunchiness, contrasting with the soft crumbly fish, to produce a classic and wholesome dish enjoyed by millions. Now my mouth is watering: I must go and cook dinner.

SIMPLE CRISP DUCK

1 duck
2 oranges, sliced

Preheat the oven to 320°F (160°C). Working from the neck, carefully loosen the skin from the meat (a rubber spatula is ideal for this task) and insert orange slices under the skin. Cook 1½ to 2 hours, depending on the size of the duck. Raise the temperature to 400°F (200°C) for the last 15 minutes in order to brown the skin.

Experiments with This Recipe

The crispiness of the skin produced in this way compares favorably with that produced by far more complicated methods, including the use of glazes. The orange keeps the skin and the meat hydrated until the final minutes of the cooking process, when the skin is crisped. Try the recipe with and without the orange. Outcomes can be assessed objectively using the texture analyzer with the acoustic envelope detector (see figure 1). Alternatively, assess by eating!

Further Reading

Chen, J. S., C. Karlsson, and M. J. W. Povey. 2005. "Acoustic Envelope Detector for Crispness Assessment of Biscuits." *Journal of Texture Studies* 36:139–156.

Szczesniak, A. S. 1988. "The Meaning of Textural Characteristics—Crispness." *Journal of Texture Studies* 9:51–59.

Varela, P., J. Chen, S. Fiszman, and M. J. W. Povey. 2007. "Crispness Assessment of Roasted Almonds by an Integrated Approach to Texture Description: Texture, Acoustics, Sensory, and Structure." *Journal of Chemometrics* 20:311–320.

three

MEDITERRANEAN SPONGE CAKE

CRISTINA DE LORENZO AND SERGIO LAGUARDA

sponge cake is not easy to make. A typical recipe calls for the beating of egg yolks and sugar until fluffy. Flour is added to the beaten yolks and sugar. Then stiffly whipped egg whites are very carefully folded into the flour mixture. All this manipulation puts at risk the airy, fluffy, and spongy character of the cake. Yet, of course, the difficulty of making a sponge cake does not detract from its popularity. The cake's airy, springy texture is most certainly the key to its broad appeal. Giving a Mediterranean twist to this classic recipe, by substituting olive oil for butter, can yield some surprisingly positive effects. Virgin olive oil is becoming an important part of a healthy diet because of its beneficial unsaturated fatty acids and antioxidants. In addition, it has wonderful flavors that tend to vary depending on the variety of olives from which the oil is extracted.

The Mediterranean culture and way of life are strongly linked to olive oil and hence the olive tree. The ancient olive tree is relatively easy to cultivate because it is not dependent on soil quality or climate. Olive oil is becoming nowadays an important focus in both science and economics. Much of the scientific activity aims to maximize the capture of this, the most highly regarded of vegetable oils. Research programs exist to understand the signature qualities of what is known as cold-pressed extra virgin olive oil (EVOO), considered the highest grade of olive oil and among the finest of all edible oils, both for its flavor and nutritional value. Also, science is enabling the discovery of new alternatives to fight the fungi and flies that are most commonly linked to declining yields among olive trees.

Olive oil of good quality has a great taste and is worth the heftier price. Different categories, or grades, exist. EVOO is the oil produced from the first pressing of olives, and the "extra" grade must be assessed by a panel of experts that is given the sole power to decide whether the olive oil is absolutely free of flavor defects. Virgin olive oil (VOO) refers also to the oil obtained after the first pressing of olives, but, in this case, the judging of the attributes is not as strict. Extra virgin olive oil must contain less than 0.8 gram per 100 grams (0.8 percent) free oleic acid.[1] It is so rare that it accounts for less than 10 percent of the olive oil produced. Like wine, EVOO varies widely in taste, color, and appearance, depending on its origin. Virgin olive oil must contain no more than 2 percent free acidity, also as oleic acid; needless to say, it is of inferior quality to EVOO. Common, or ordinary, olive oil should contain no more than 3.3 percent free acidity, and defects are frequently found in this grade of oil. We chose to highlight the olive oil in this recipe, given that it is the only culinary fat among the ingredients. To this end, we used a monovarietal EVOO from Madrid, extracted from ripe cornicabra olives, which are characterized by their piquant and medium-bitter taste.

The classic sponge cake is a mixture of nearly equal parts flour, sugar, eggs, and butter, each of which has a role to play in the development of the textural and sensorial elements characteristic of the sponge. However, eggs are probably the most important. This is because the beaten egg whites provide most of the bubbles that, during the early stages of baking, expand and lend the cake its airy and spongy texture. As baking progresses, the starch in the flour cooks and supplies strength against collapse—this is why flour is an indispensable ingredient in every cake recipe.

With some context provided, we offer a recipe for the modern time-saving kitchen that provides a taste of the ages-old healthy Mediterranean diet. We chose to make sponge cakes flavored with EVOO. In addition, we wondered if there was a way to further enhance the sponginess of the sponge; that is, we sought to find a way to maxi-

1 Oleic acid is a fatty acid. Fatty acids are the building blocks of triglycerides, which are formed by attaching three fatty acids to a glycerol molecule. Triglycerides are the constituents of fats and oils. When a fatty acid is cleaved from a triglyceride chain, it is referred to as a free fatty acid. These are associated with the development of off flavors in fats and oils; hence, the lower their content, the better.

OLIVE OIL SPONGE CAKE

½ ounce (15 g) all-purpose (wheat) flour
1¾ ounces (50 g) sugar
4 large eggs
3½ ounces (100 mL) extra virgin olive oil
0.7 g xanthan gum
2 charges nitrous oxide (for a 1-quart [1-L] syphon)

Preheat the oven to 375°F (180°C). Accurately weigh all the ingredients. Sift the flour and add the sugar, eggs, and olive oil. Mix thoroughly with the aid of a mixer. Scatter the xanthan gum and mix thoroughly until the dough becomes thick and homogeneous. Fill the syphon with the batter and charge it with the nitrous oxide. Let the dough stand in the syphon for 15 minutes. Shake well before dispensing. Fill muffin cups from one-half to three-quarters full and bake for 15 minutes. Olive powder (made from defatted, dried olives) may be added to the dough before baking for an extra flavor kick.*

* For the olive powder: take 3½ ounces (100 g) of green or ripe olives, as preferred. Pit (de-stone) the olives if necessary. Cut in little pieces and reserve them in a 1 percent solution of ascorbic acid; prepare flat on a baking sheet and dry at 190°F (90°C). When dried, scatter a little ascorbic acid and grind to obtain a fine powder.

mize its final volume. One of the easiest ways to do this is to place the batter in a whipped-cream dispenser, or syphon, which is then pressurized with nitrous oxide. This effectively introduces thousands of tiny bubbles into the batter without the need for beating. With the use of the dispenser, we hypothesized, the final volume, smoothness, and lightness of the sponge would be maximized. We introduced xanthan gum, a natural thickening agent, to help prevent gas escaping from the sponge. Ingredients like xanthan gum are known to influence dough expansion and stability by increasing the viscosity and strength of the thin films of dough that surround the air bubbles (for a detailed account of what xanthan is and how it works, see chapter 19). The sponge cake recipe we used for our experiments was modified to make the eggs the main structure-building ingredient.

Our syphoned olive oil sponge cakes were highly successful among tasters, who acknowledged their smoothness and, most important, their sponginess. As a control, we prepared olive oil sponge cake in a "traditional" way, without the syphoning, and, as a test of xanthan gum's properties, we made a second syphoned sponge, omitting this ingredient. We were careful to keep everything else the same across the three recipes.

Table 1 describes some of the measurements we made. It is obvious that the traditional sponge was taller than the others. In the syphoned sponge without xanthan gum, there was a heterogeneity in the size and distribution of air within the cake, as revealed in

Table 1 Physical Properties of Three Olive Oil Sponge Cakes

	Traditional sponge cake	Syphoned sponge cake	Syphoned sponge cake + xanthan
Height	2⅜ in. (6.0 cm)	1⅞ in. (4.8 cm)	1½ in. (3.8 cm)
Diameter		2⅝ in. (6.5 cm)	
Flattening in height	0%	40–70%	less than 20%
Total porosity	10.4%	**12.9%**	11.4%
Porosity (% voids smaller than 1 mm²)	0.92%	1.19%	**1.48%**

Note: Bold figures indicate statistically significant differences. Measures are mean values from image analysis of three independent photographs.

the cross-section image (figure 3). A possible explanation is that the sponge contained too many bubbles to begin with. This meant that a rapid volume expansion took place too early during baking, resulting in a small number of large bubbles that the batter was unable to contain. In the case of the syphoned sponge with added xanthan, the individual, somewhat distorted, bubbles were still distinguishable; that is, the gum prevented them from merging. The presence of xanthan gum increased the viscosity and strength of the batter, which might have been high enough to limit both the incorporation of nitrous oxide into the batter and its expansion during baking. A more

Figure 3 Cross sections of three different sponge cakes that illustrate the nature of the crumb: traditional (*left*), syphoned without xanthan gum (*center*), syphoned with xanthan gum (*right*).

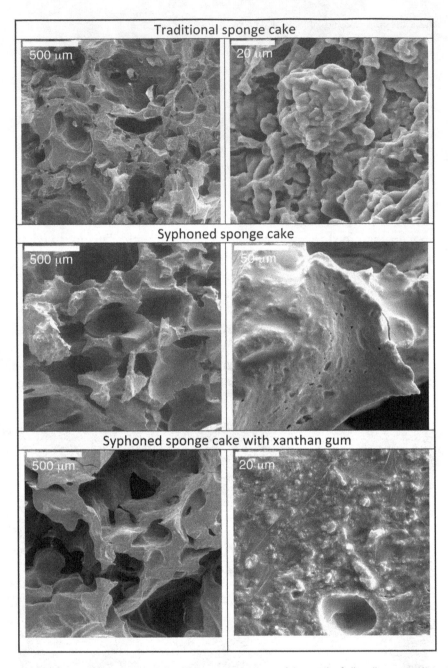

Figure 4 Cryoscanning microscopy, at different magnifications, of the crumb of olive oil sponge cakes made in the traditional way and with a whipped-cream dispenser in the absence and presence of xanthan gum.

viscous batter means that the bubbles must push harder to increase the volume of the cake.

Despite having the lowest volume of all the cakes tested, the syphoned sponge cakes with added xanthan were the most porous—offering the best indication of sponginess—as evidenced by the large number of voids, or spaces, within the crumb. As the *qualitative* evidence of our culinary findings, we present microscopic images of the crumb structure of the sponges. It is common to find answers to the *macroscopic* behavior of foods by looking really, really closely into their *microstructure*. The use of advanced scanning microscopes gave us a view into the core of the cakes. Figure 4 shows the microstructural details of the three different sponges (traditional, syphoned, and syphoned with xanthan).

It seems as if the voids are the result of a series of explosions. In fact, they are. They represent the remains of bubbles once embedded in the sponge. When xanthan gum is added to the syphoned batter, the general structure is comparable to that of the syphoned sponge *without* xanthan. However, the cake is much more aerated, as evidenced by the presence of overall smaller voids and an apparently elastic and membrane-like structure. And the degree of aeration helps us explain the differences observed during the eating of the sponges.

How does this affect the texture of the sponge cakes? Texture is one of the most important and characteristic properties of foodstuffs. It is no wonder there are several chapters in this book that touch on it (for example, chapters 18 and 23). To assess texture, we need to measure it. There are at least two ways to measure texture: one is by instrumental analysis (see, for example, chapters 2 and 21), and the other is by means of our own biological instruments—our mouths. Not surprisingly, trying to correlate instrumental measures with sensory perception is very difficult; in fact, scientists are always struggling to correlate these measurements, normally having to design specific tests for each specific food under study. Of course, not all foods are created equal. In table 2, we present a list of some of the relevant textural properties that we believe describe the quality of the olive oil sponge cake—that is, its texture profile. This profile includes properties such as *hardness*, a measure of the cake's "bite"; *cohesiveness*, the cake's ability to hold together (for example, a crumbly texture is not cohesive); *adhesiveness*, the cake's tendency to stick to your teeth or palate; and *springiness*, the speed and degree of the cake's recovery after being bitten into.

The numbers shown in table 2 mean that the traditional sponge cake was the hardest, the least cohesive, and the most adhesive, while

Table 2 Textural Properties of Three Olive Oil Sponge Cakes

Test property/sample	Texture profile analysis		
	Traditional sponge cake	Syphoned sponge cake	Syphoned sponge cake + xanthan
Hardness	598	434	122
Cohesiveness	0.36	0.45	0.55
Adhesiveness	69.8	52.7	50.0
Springiness	0.45	0.23	0.67

remaining somewhat springy. The syphoned sponge with xanthan gum was the least hard and the most springy, which we believe was related to the sponge's increased porosity (degree of voids from bubbling).

These *instrumental* results were in general accordance with our tasting panel's findings. The sensory analysis came from ten trained volunteer tasting judges. The tasters reported that the traditional (control) sponge was the hardest, firmest, and driest; this sponge was considered to be neither adhesive nor greasy. It was basically neutral on the tongue but with noticeable EVOO notes. The syphoned sponge cakes without xanthan were spongier than the control and had a similar EVOO flavor. Approximately 50 percent of the tasters expressed the perception of a greasy sensation. The syphoned xanthan gum cakes were the softest and most springy. They were also considered neutral to the tongue but nicely oily with the distinct flavor of olive oil—the clear winners.

It is interesting to note that that the syphoned cakes, especially those with xanthan gum, were extremely easy to deglute, or swallow.

The manipulation of texture—a very complex and multidimensional property—by exaggeration of the proportions of certain ingredients (in this case, the volume of air in the sponge cake) can be a rewarding experience. Insights into the variables that make this kind of manipulation a success, such as the increase in viscosity by the addition of xanthan gum, reveal how far a little extra knowledge can take us in our culinary adventures.

Further Reading

Visser, Margaret. 1986. "Olive Oil: A Tree and Its Fruits." In *Much Depends on Dinner*, 224–258. New York: Grove Press.

four

SPHERIFICATION

Faux Caviar and Skinless Ravioli

CÉSAR VEGA AND PERE CASTELLS

water immobilization is a cool thing! The simplest way to accomplish it is by freezing. But can you think of how water might be immobilized (so to speak) at temperatures above freezing, say at 50°F (10°C)? Think Jell-O and a new process that mimics caviar and you have two methods that nearly stop water in its tracks.

Scientists refer to the phenomenon that creates Jell-O as *gelation* and to the process that produces mouth-popping liquid-filled beads as *spherification* (much more on that in a moment). To understand gelation, visualize water molecules being trapped in time and space— they can no longer move, and the food that contains these molecules exhibits solid-like behavior. The food does not flow; it has become a gel.

For this to happen, a gelling agent is needed.

Most gelling agents, commonly known as gums, are naturally occurring macromolecules; that is, their molecular size is thousands, or millions, of times larger than, say, a single sugar molecule. In fact, scientists refer to gelling agents as polysaccharides because they are very long chains of sugar molecules linked to one another. Once they are dispersed in water, the final mix will behave in a concentration-dependent fashion: at low concentrations, their molecules have minimal interaction, and this results in an increase in viscosity. Viscosity refers to how easily a liquid flows. The harder it is to accomplish flow, the more viscous the liquid. For example, honey is more viscous than water (for a detailed description of the viscosity of many common foods, see chapter 18). There are molecules that when added to water increase its viscosity; as the concentration increases and under certain conditions (cooling, the presence of certain ions, or change in the acidity, for example), the molecules associate with one another so that the result is a tridimensional structure that acts less like a liquid and more like a solid.

Gelatin is the oldest of gelling agents and today remains a popular ingredient in our pantry. We all know that gelatin is the gelling agent in a packet of Jell-O that causes the liquid blend to solidify upon cooling, after the mix is dissolved in water and brought to a boil. But is it possible to make gels without the bother of heating or cooling them? The short answer is yes. A variety of relatively new gelling agents allow the culinary practitioner to create new textural experiences previously unknown in the context of the kitchen.

As chefs have started to travel across continents in search of inspiration and the world has become more global, so has the kitchen. This has led to the incorporation of ingredients from all over and to the creation of what we now know as fusion cuisine. The first important example of this fusion, dating back to the 1970s, was nouvelle cuisine, which was the intermingling of Asian ingredients and techniques, emphasizing lightness and clarity of flavor, with the French culinary practice.

Chefs have not stopped traveling since. In 1998, El Bulli, the world-renowned restaurant located in the Spanish town of Roses on the Costa Brava, introduced agar-agar, a natural seaweed extract that serves as the vegetarian counterpart to gelatin (which is derived from animal tissue). It is a very common ingredient in Asian cuisine, where it is used as a gelling agent in the preparation of jellies, puddings, and custards. What is unique to agar-agar gels is their ability to resist melting to 140°F (60°C) and that they show less slippery and less brittle mouthfeel than gelatin-based gels. Imagine the possibilities! The rediscovery of agar-agar has led to the creation of dishes such as parmesan spaghetti, hot lobster gelatin, sweet potato jelly, and vinaigrette sheets, among many others that have changed the paradigm of what a culinary gel can be.

Another revolution was the spherification technique. In its simplest form, spherification is a culinary technique in which a gel forms

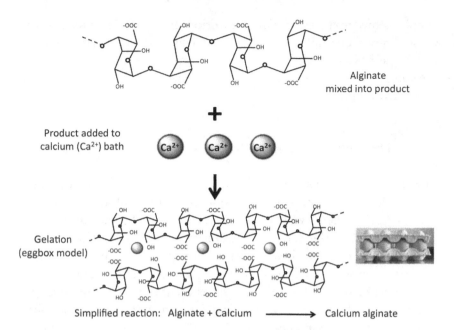

Alginate mixed into product

Product added to calcium (Ca²⁺) bath

Gelation (eggbox model)

Simplified reaction: Alginate + Calcium ⟶ Calcium alginate

Figure 5 Schematic representation of the gelation mechanism of sodium alginate in the presence of calcium ions (Ca²⁺). Calcium (or any other divalent ion) acts as a bridge among the polymeric chains of alginate and fits among them like eggs in an eggbox. This eggbox configuration effectively links the chains together by forming connections. When sufficient connections are formed, the product gels, and the skin of the sphere is formed.

around a liquid center, like caviar. In this case, the gelling agent is sodium alginate, which, like agar-agar, is extracted from seaweed. The spheres are created through a calcium-mediated gelling mechanism. First, sodium alginate is dispersed into the solution that the spheres will be made of, for example, apple juice; this solution is then loaded into a syringe and carefully dripped into a calcium chloride bath. As soon as the alginate-containing solution comes into contact with the calcium bath, the solution gels, creating a sphere with a liquid center!

This happens because calcium acts like a bridge between chains of alginate, enhancing their interactions and favoring gelation. Scientists often explain this gelation mechanism using the eggbox model (figure 5). Interestingly enough, this model only works well with calcium. Other divalent atoms, like Mg^{2+} (magnesium), create weak gels. Another virtue of these types of gels is that they are not thermoreversible; that is, they do not melt on heating.

From a culinary standpoint, the advent of this technique represented a major step forward because it allowed the creation of two simultaneous textures: a *liquid* interior and a *solid* exterior—the most famous examples being apple caviar and pea ravioli. As the technique has evolved, the diffusion of gas (for example, CO_2) into the sphere made it possible to develop three-phase drinks, such as the spherified mojito!

A 360-Degree Look at Spherification

In 2003, Pere Castells dined at El Bulli, where he had his first rendez-vous with the spherification technique in the form of apple "caviar." Ferran Adrià, celebrity chef and owner of El Bulli, had no problem showing Castells the actual "manufacturing" process. Castells was stunned to witness a kitchen preparation procedure making use of scientific concepts. In fact, Castells, a trained chemist and teacher, had included some of Adrià's previous creations in his high school chemistry textbook upon witnessing the tremendous power of illustrating science through cooking. Castells's profile and keen culinary interest prompted Adrià's curiosity regarding the relationship between science and cooking.

As the two men got to know each other, and with Castells's expertise in chemistry, they resolved some of spherification's technical issues. Nowadays, spherification is ubiquitous in restaurant kitchens. However, the technique is not as simple as merely dropping alginate-containing liquid into a calcium chloride bath. Spherification has its challenges.

- *Get the acidity right.* During the making of apple caviar, mixing sodium alginate with apple juice, at levels that allowed skin formation, rendered a viscous liquid center. Adrià and his team aimed for something more fluid so as to exaggerate the difference between the outer shell and the liquid interior. After careful consideration of the range of possibilities, the acidity of the apple juice seemed like a good place to start experimenting. With the help of universal pH indicator strips, it was found that the alginate–apple juice mix was indeed very acidic (with a pH of about 3.5), which was not surprising, given that they were using tart Granny Smith apples.[1] Castells hypothesized that this acidic environment would favor the creation

---

1 For reference, the pH of lemon juice is about 2.4; that of milk, about 6.5; and that of egg white, about 9.0.

of alginic acid, an insoluble compound that would thicken the solution. Adrià and his team happened to have trisodium citrate in their kitchen pantry—who knows why! Trisodium citrate is a weak alkali derived from citric acid that can increase the pH of slightly acidic concoctions, while also trapping some of the calcium ions in the bath (if there is an excess of calcium, the liquid inside the spheres tend to be very thick). They tested it and voilà! The blend became significantly less viscous! This finding enabled the design of mango ravioli—a creation that plays a game with your senses. These ravioli look and feel like raw egg yolks but taste like pure mango goodness!

• *Get the calcium right.* As the team kept pushing the boundaries of this new application, they ventured into using spherification with dairy products. To their surprise, upon the addition of alginate to a milk-based sauce, for example, the mix gelled instantly and irregularly (imagine the strands of alginate as those formed by egg whites in egg-drop soup). It did not take long to figure out that the calcium present in the dairy was responsible for what they observed. Did this mean that spherification was not possible with milk or yogurt? Not necessarily. Castells and Adrià investigated further.

• *The life of a single droplet.* If we were to place ourselves inside a drop, or sphere, of alginate-containing apple juice as it is submerged into the calcium chloride bath, we would observe that the gelation process is not static. Upon contact, calcium and alginate form a permeable film around the sphere. Such permeability means

SPHERICAL MANGO RAVIOLI

8½ ounces (¼ L) water
1.3 g sodium citrate
1.8 g sodium alginate
8½ ounces (250 g) mango puree

Gelling Bath

1 quart (1 L) water
5 g calcium chloride

Combine the sodium citrate with the water. Add to this the sodium alginate and blend until well incorporated. Bring to a boil. After the mixture cools down to room temperature, add the mango puree. In a different container (1 to 1½ inches deep), mix the water with the calcium chloride. Add the mango puree mix to the calcium chloride bath a spoonful at a time. The size of the ravioli will be defined by the size and shape of your spoon. Leave the spheres in the bath for 2 minutes. Then, with the help of a slotted spoon, move the spheres to a container full of cold water and rinse. Serve within 10 minutes.

Adapted from www.texturaselbulli.com

that transport of small enough molecules across the film can occur—commonly known as osmosis. This explains why a sphere of sweetened apple juice placed in water can eventually lose its sweetness as sugar migrates out of the sphere. The same occurs with calcium ions, but in this case, as they move from the bath into the sphere, they eventually promote complete gelation of the sphere—not the desired effect. The problem is exacerbated by the fact that, no matter how well the spheres are rinsed (to remove calcium chloride from their surfaces), calcium continues to penetrate into the spheres. In practical terms, this means that the "life span" of our spherical ravioli is, at most, ten minutes, after which, it forms a very thick and unappealing skin, and the sensation of an "explosion" in the mouth is either decreased or lost. Faux caviar lasts only about five minutes after it is made.

It became the team members' obsession to solve this problem. They tried to eliminate excess calcium and therefore prevent its diffusion throughout the sphere. This proceeded without much success. The attempts either sequestered too much calcium, which inhibited gelation, or sequestered too little to delay or slow down its diffusion. The team finally decided to reverse the process and have the calcium moving outward.

Calcium chloride was added to peach juice, and this mix was then dripped into a bath of sodium alginate: it was the birth of "inverse spherification." Castells still remembers the exact date—February 20, 2005.

In essence, the gelation process is the same, but in this case, the calcium in the product moves outward *and* is available only in limited amounts (whatever is contained in a single droplet). Therefore, the center always remains liquid, which, by the way, also solves the issue of making spheres with dairy products!

• *Ingredient selection and flavor balance.* Many applications were possible, but there was still another problem: calcium chloride has an unpleasant and hard-to-mask taste. The solution to this problem comes as a mixture of calcium gluconate and lactate (that is, calcium gluconolactate), which gives a nearly flavorless result.

• *Solution density/viscosity balance.* In the case of inverse spherification, it was rather common to find that the droplets of some liquid preparations did not sink into the alginate bath and therefore inhibited the formation of the spheres. This was mainly because the density and/or viscosity of the bath was much higher than that of the falling liquid droplets. As a means to alleviate this difference, xanthan gum

Figure 6 Green-olive spheres were created by making a fine puree from green oils to which a calcium source was added. The puree was then carefully spooned into an alginate bath, where an instant skin was formed.

was added. By use of this technique, the creation of the liquid olive was made possible (figure 6).

To summarize, successfully carrying out spherification means paying attention to a series of factors: the acidity of the liquid, the amount of calcium already present, where the calcium is (that is, in the bath or in the droplets), and how the calcium is sourced (calcium chloride or calcium gluconolactate). Last but certainly not least are flavor and viscosity.

Most everyone agrees that there is value in approaching problem solving in a systematic and multidisciplinary way, no matter the field of study. Strangely enough, when the problem revolves around food, everything seems to go awry, despite the systematic approach. Why? We still wonder.

It has taken decades for the culinary world to accept as natural the current dialogue between science and the kitchen. The fact that this conversation was so long in coming reveals the reluctance and arrogance of both the scientist and the cook. It might also say something about the barriers language poses. However, cooks and scientists are now talking to one another, and this dialogue has already borne fruit. Science brings control, precision, and consistency to the kitchen, which can mean only one thing: better cooking!

Further Reading

Andrews, Colman. 2010. *Ferran: The Inside Story of El Bulli and the Man Who Reinvented Food*. New York: Gotham Books.
Lersch, Martin, ed. 2010. "Texture: A Hydrocolloid Recipe Collection." Available at http://khymos.org/recipe-collection.php.

five

KONJAC DONDURMA

Designing a Sustainable and Stretchable
"Fox Testicle" Ice Cream

ARIELLE JOHNSON, KENT KIRSHENBAUM, AND ANNE E. MCBRIDE

around the world, culinary artisans and innovators alike use myriad ways to manipulate sweet frozen cream. Different preparations yield distinct flavors. But what can perhaps best distinguish one ice cream type from another is its texture. From granita to gelato, ice cream can take on a range of traits, from granular to velvety. Ice cream producers seek to create a deliciously flavored, creamy, and pliable cold dessert by freeze-thickening a liquid until it reaches a semisolid state. Churning a sweetened cream mixture as it freezes prevents large ice crystals from forming, as in Philadelphia-style ice cream. The addition of egg yolks, as in gelato or French-style ice creams, results in a denser, richer, more custard-like ice cream.

Ancient recipes can give ice creams unique and exotic textures that evoke the current experimental trends of avant-garde pastry chefs. One such inspiration for the modern kitchen, salep dondurma, is a Turkish frozen preparation that can become so thick and chewy as to require cutting with scissors.

Salep dondurma is traditionally made with sweetened goat's milk and salep flour, a powder ground from the roots of the *Orchis mascula* (an orchid indigenous to Anatolia, Turkey). The roots of the plant are called *salep*, a name derived from the Turkish word for "fox testicle," which describes their appearance and alludes to their putative aphrodisiac and virility-enhancing qualities. Salep dondurma

Figure 7 Stretching salep and konjac dondurma.

literally means "fox testicle ice cream." Its presumed birthplace is the Turkish city of Kahramanmaraş, providing another name for the dish: Maraş dondurma. Salep dondurma likely derives from an ancient beverage, a pudding-like warm drink of milk thickened with salep flour, also called salep (or sahlab). Salep has been consumed around the Mediterranean for a millennium and was popular in places as distant as England in the eighteenth century, before being supplanted by coffee or tea as the morning beverage of choice. The salep flour thickens the hot milk, giving a robust body and a somewhat gooey mouthfeel. When the drink is frozen, it becomes a uniquely pliable and chewy ice cream that can be pulled and stretched into a long, thick rope, like cold, soft taffy (figure 7).

On a physiochemical level, ice cream is a frozen foam composed of air (introduced by churning, whipping, or kneading during freezing); small ice crystals; partially coalesced globules of milk fat that stabilize the air bubbles by surrounding them; and water, milk proteins, and sugar. In creamier varieties of ice cream, large protein molecules (from the milk) and polysaccharide molecules (long chains of linked sugar units) yield a thicker texture and smoother mouthfeel by preventing as many ice crystals from growing or by increasing the viscosity of the unfrozen phase (for more on the science of ice cream, see chapter 17).

In Turkey, adept and playful salep dondurma sellers turn this street food into theater, using long, stiff poles and paddles to lift, stretch, and flip their ice cream out of their freezer carts. One of their tricks is to place a serving of dondurma into a cone held by a customer and then immediately yank the pole back, extracting the ice cream from the cone in one swift motion (see "Turkish ice cream: Maraş dondurmasi" [a]). Kneading and stretching the salep dondurma has a functional benefit: it develops and maintains the ice cream's stretchiness, similar to the kneading of bread dough (see "Turkish ice cream: Maraş dondurmasi" [b]). Dondurma's texture results not only from the dramatic exertions of the vendors but also from its unusual ingredients—which is what prompted our study of salep dondurma's stretchiness at the molecular level.

The orchid tubers used to make salep flour contain significant amounts (about 55 percent on a dry basis) of the polysaccharide carbohydrate salep glucomannan, composed of glucose and mannose sugar units linked together through strong chemical bonds. Glucomannan molecules are long and chainlike, referred to as polysaccharides, and can contain more than 5,000 sugar units each, resulting in highly extended chemical structures (figure 8). Glucomannans, like most carbohydrates, are hydrophilic, which means that they can establish many attractive interactions with water. Glucomannans and other similar "hydrocolloid" molecules can form entangled networks between polymer chains when added to an aqueous solution. They are sometimes used in ice cream to control ice crystal growth or provide enhanced viscosity during melting.

As illustrated in chapter 4, the addition of certain molecules to water can yield a gel—a semisolid network of polymers and liquid water. The viscoelasticity of gels may differ depending on processing and on the presence of other components, such as sucrose or milk. The large molecular weight of glucomannan, combined with its attractive interactions with water and other molecules in ice cream, gives rise to the highly viscoelastic properties that are critical to the texture and extensibility of salep dondurma.

Today, it is nearly impossible to obtain, outside of Turkey, authentic salep flour, the key ingredient in salep dondurma, because the Turkish government has restricted its export in response to reports of declining orchid populations. Foreign-produced

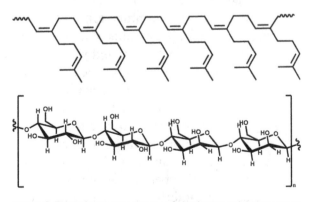

Figure 8 The polymer components of dondurma: poly-beta-myrcene (*top*) and glucomannan (*bottom*). These molecules may be 5,000 or more units long.

dondurma often includes common starches, such as cornstarch or arrowroot, and lacks the physical attributes of genuine Maraş dondurma. We developed a scientific interest in stretchy ice cream, allied with a desire to produce it conveniently and inexpensively. Our goal was to retain the unique texture of the original dish with an eye toward *Orchis mascula*'s sustainability, which prompted us to seek possible substitutions for salep flour. Initial attempts, using cornstarch and arrowroot, produced a thick but unstretchy ice cream. We soon realized that any alternative would ideally possess a molecular composition similar to that of salep flour, leading us to consider konjac flour. Like salep, konjac flour (also known as konjac mannan) is rich in glucomannan and manufactured from the bulb of a tuber that, like *Orchis mascula*, is also suggestively named: *Amorphophallus konjac*—the genus *Amorphophallus* means "shapeless penis" in Latin.

Konjac flour is commonly used in Japanese cuisine to produce a kind of gel called *konnyaku*, which can be made into noodles called *shirataki*. To our knowledge, konjac mannan is not used as an ice cream additive. Konjac flour is plentiful, relatively inexpensive, and is not derived from a wild endangered plant. We discovered that konjac flour can be used to produce a stretchy, chewy ice cream whose only other ingredients are milk, sugar, and optionally, mastic resin, a traditional additive. In homage to its Turkish inspiration, we call this dessert konjac dondurma.

Harvested from *Pistacia lentiscus* trees, which grow on the southern half of the island of Chios, located in the eastern Aegean Sea, mastic has been used for food and medicinal purposes since antiquity. It was described in writings by Pliny and Dioscorides nearly two millennia ago and was also identified in Egyptian mummy balms dating back to the seventh century B.C.E. The sap of the *Pistacia lentiscus* tree contains large amounts of hydrocarbon-based compounds known as terpenes, which are found widely in nature in different compositions of five-carbon building blocks and play an important role in, among other things, the floral, spicy, and herbal aromas in wine. One of the compounds found in *Pistacia lentiscus*, beta-myrcene, is believed to polymerize as the sap runs from the tree and forms poly-beta-myrcene, the major polymer component of mastic resin (see figure 8). Not all salep dondurma recipes call for mastic, so its role in dondurma is currently open for debate. Some people suggest that mastic is only a flavor additive, providing no functional role. Mastic has a strong pinelike aromatic note, and during preliminary testing, we found its addition to ice cream seemed to amplify its thickness and pliability.

Further experiments are necessary to evaluate the potential role of mastic in defining the texture of dondurma.

Methodology and Results

We sought to conduct an initial evaluation of the roles of salep, konjac, and mastic in the texture of dondurma, starting with a reliably reproducible recipe. A general method for creating a glucomannan-thickened, stretchy ice cream requires several steps:

- The suspension in milk of the chosen glucomannan, either salep flour or konjac, as well as mastic if desired
- The addition of sugar
- Boiling, to concentrate and promote hydration of the glucomannan
- Beating the cooked mix, for effective sugar and glucomannan dispersion
- Freezing
- Beating and kneading the frozen ice cream, to develop chewiness and stretchiness

Before the advent of refrigeration, salep dondurma was likely frozen in an ice-brine solution. We tried this in our laboratory kitchen, which made for an extremely laborious process. Many large dondurma-producing companies in Turkey reportedly use gelato makers to freeze their ice cream, after which it is hand-kneaded with rods—a technique we found to be highly effective. We also tried liquid nitrogen to directly freeze the dondurma in a stand mixer, which produced similarly stretchy, if slightly less dense, results compared with dondurma created in an ice cream maker.

A standard salep dondurma recipe calls for a concentration of 0.8 percent salep flour and 22 percent sugar. The milk is heated first, then sugar and salep are added, and the mixture is boiled. For konjac dondurma, we initially used the same proportion of konjac—that is, 0.8 percent, which resulted in a sticky mouthfeel. Therefore, we gradually reduced the amount of konjac flour to 0.4 percent, which produced a texture closer to that of characteristic salep dondurma. A variation on this procedure involved suspending the konjac flour in the milk before heating, as the addition of konjac to hot milk caused it to clump and burn. We also tried adding mastic to the recipe in the amount of 0.5 to

KONJAC DONDURMA

1½ quarts (1.6 L) whole milk
14½ ounces (410 g) sugar
8 g konjac flour (Nutricol GP 312)*
2 g mastic (Chios gum mastic large tears), optional†
2 quarts (2 L) liquid nitrogen, plus more for mastic if
    using (or dry ice if preferred)

If using dry ice, chill the mastic over the solid CO_2 for 5 to 10 minutes or freeze it instantly with liquid nitrogen, then grind it into a fine powder in a chilled mortar and pestle.

Slowly whisk the konjac flour into the milk until it is completely incorporated. Set aside for 30 minutes, until the mixture thickens.

With a candy thermometer handy, pour the mixture into a 4-quart (4 L) pot and heat over medium-high heat. When the mixture reaches 125°F (50°C), whisk in the sugar and the mastic. Bring the mixture to a boil. The mixture will increase in volume; keep boiling at this speed, whisking constantly (with care not to let it boil over), for 15 minutes.

Pour the mixture into the bowl of an electric stand mixer fitted with the paddle attachment and beat for 30 minutes, until cool. Slowly pour in the 2 quarts (2 L) of liquid nitrogen, 3½ ounces (100 mL) at a time, letting the mixer run 1 to 2 minutes between additions, until the cloud of vapor has cleared. The ice cream will reach its optimal texture at 10°F (−12°C).† Consume immediately, or store in the freezer for up to two months.

Yield: about 2 quarts

* Nutricol GP 312 konjac flour was made available by FMC BioPolymer (http://www.fmcbiopolymer.com).

† Large tears of chios gum (or mastic) are available at the Mastihashop in New York and at http://www.mastihashop.com.

† Once the ice cream reaches 10°F (−12°C), further beating with a wooden spoon or in an electric mixer fitted with a dough hook will increase its stretchiness and chewiness. This is best done for a few minutes at a time, followed by 10 to 20 minutes in the freezer to prevent melting. If liquid nitrogen is not available, freezing can be achieved by placing the mix for 30 to 45 minutes in an ice cream maker that has either a frozen bowl or a built-in compressor. Once the ice cream is frozen, beat it by hand to develop stretchiness and chewiness.

2.5 grams per liter of mix. We noted very large flavor differences, depending on the mastic dose, and generally more pliability and chewiness as the amount of mastic increased. Additional controlled testing would be required to establish quantitative results in the area of texture.

Besides tasting many dondurma variations, we gathered an informal sensory panel composed of ice cream lovers, Turkish expatriates, and other interested parties. We asked individuals to taste both salep and konjac dondurmas and compare them for taste, chewiness, and stretchiness. These side-by-side comparisons of physical attributes and flavor revealed that konjac dondurma matches the properties of salep dondurma: it stretches and resists cutting in much the same way and has the same pillowy, chewy mouthfeel as the original. Testers familiar with dondurma as produced in Turkey confirmed that the stretchiness and chewiness of konjac dondurma resembled the Turkish original.

Viscoelasticity is the ability of a food to exhibit

both viscous (liquid) and elastic (solid) behavior. Stretchiness, a dimension of viscoelastic behavior, refers to the ability of a food to be pulled from its ends in opposite directions without breaking. It is a property that is difficult to measure in a reproducible way (even more so in frozen systems). We tried to compare the relative stretchiness of dondurma made with salep to that made with konjac. We did not find a significant difference between them. However, we believe that scientists should examine salep and konjac more closely—it would surely prove to be a mind-stretching investigation!

The fascinating culinary attributes of dondurma are due to, we believe, the unusual molecular composition of salep flour, which contains a large amount of the polysaccharide glucomannan. Konjac dondurma seems to replicate the properties of the original salep dondurma, as we have shown. The successful use of konjac as an alternative glucomannan source for stretchy ice cream implies that this traditional treat can be produced using ingredients that are both readily available and sustainable.

Our experience making dozens of batches of dondurma indicates that processing is essential in developing stretchiness. Failing to beat dondurma (either salep or konjac) in the unfrozen state or to knead and stretch it during freezing tends to lead to a grainy, nonchewy, poorly cohesive ice cream that will not stretch.

We can imagine additional gastronomic challenges. If the conditions required for stretchiness in this food were more completely understood, perhaps other innovative textures could be produced in other foods. It remains to be established whether it is critical to include denatured proteins to obtain extensibility. Also undetermined is the role that mechanical action plays in establishing dondurma's stretchy and chewy properties. Further experimentation is required to elucidate these parameters individually and in combination. How else might glucomannans be used in food with delicious and exotic effects? A topic to stretch the imagination—and to chew on.

Acknowledgments

We gratefully acknowledge the assistance provided by science and food writer Patricia Gadsby; Yeliz Utku, Ph.D.; Ricky Silver; Will Goldfarb, chef-owner of WillPowder; David Arnold, director of technology at the French Culinary Institute, New York; Amy Bentley and Kaitlin Goalen, New York University; Johnny Iuzzini, executive pastry chef at Jean-Georges, New York; Unilever and its employees Ting-An Huang and Einav Gefen; Noel Kirshenbaum; FMC corporation; Mastihashop and Artemis Kohas; Tom Pold; and Ron McBride. We extend our thanks to the National Science Foundation for support through CAREER Award 0645361 and to New York University for its support through a research challenge grant.

Further Reading

Chase, Holly. 1994. "Suspect Salep." In *Look and Feel: Proceedings of the Oxford Symposium on Food and Cookery 1993*, edited by Harlan Walker, 44–47. Totnes, Eng.: Prospect Books.

Kaya, S., and A. R. Tekin. 2001. "The Effect of Salep Content on the Rheological Characteristics of a Typical Ice-Cream Mix." *Journal of Food Engineering* 47, no. 1:59–62.

McGee, Harold. 2007. "The Curious Cook: Ice Cream That's a Stretch." *New York Times*, August 1.

Tamer, C. E., B. Karaman, and O. U. Copur. 2006. "A Traditional Turkish Beverage: Salep." *Food Reviews International* 22, no. 1:43–50.

"Turkish ice cream: Maraş dondurmasi" (a). Available at http://www.youtube.com /watch?v=oisweeaQKEY&.

—— (b). Available at http://www.youtube.com/watch?v=Scbzud540vM&.

six

STRETCHY TEXTURES IN THE KITCHEN

Insights from Salep Dondurma

TIM J. FOSTER

as nicely described in chapter 5, salep dondurma is a fascinating Turkish ice cream. It has an unusual thick and stretchy consistency, which makes it chewy—and yet it also smoothly melts in the mouth. This chapter explores the molecular origins of this stretchy texture and postulates how such textures can be produced in the kitchen using alternative ingredients away from the restrictive confines of ice cream machines.

Novelty in food production, either on a commercial scale or in the home kitchen, often involves changes in product texture, as this is one of the factors controlling the organoleptic quality (taste, odor, color, and feel) of foods. From a consumer perspective, the texture provides an indication of whether a product has a too thick or too thin mouthfeel. Researchers have investigated the molecular properties of salep dondurma. Knowing these properties helps explain the relationship between salep's structure and its effect on texture. As explained in chapter 5, the main component of salep is a glucomannan (a polysaccharide containing both glucose and mannose). It is similar to that found in the tubers of *Amorphophallus konjac*, a plant that is widespread in tropical East Asia and is used as source for konjac glucomannan—the sustainable alternative tuber flour employed in making salep dondurma. Konjac glucomannan is used extensively in Japanese cuisine in varying forms, as slabs or balls of gel, as chewy noodles, and as additives in savory condiments, such as soy sauce or mustard, and in weight-control powders that hydrate to form thick, stomach-filling preparations.

The properties of salep in solution are similar to konjac glucomannan and also to certain food thickeners, such as guar gum and locust bean gum, which are generally known as galactomannans. Locust bean gum is known to gel during freezing in the production of ice cream, while salep does not (similar to guar gum and konjac glucomannan). This suggests that the properties of salep, an exotic and increasingly rare ingredient, can be replaced by more conventional additives.

Both guar gum and locust bean gum phase-separate when mixed with milk. Such phase separation is explained by the concept of incompatibility. That is, the polysaccharides and milk proteins occupy their own phase when cosolubilized in water, rather than sharing the available volume as one phase. These incompatible mixtures are known as water-in-water emulsions. These types of emulsions have properties that are similar to incompatible mixtures more often used in the kitchen. Examples of incompatible mixtures are oil-in-water emulsions, such as vinaigrette dressings, mayonnaise, and table spreads.

The phase separation in incompatible mixtures is dependent on the starting composition. Changing the composition will not only change the volume of the separated phases but also the structure of the phase-separated system. As depicted in figure 9, adding more polysaccharide (for instance, locust bean gum) increases the relative volume of the polysaccharide, or black, phase and decreases the relative volume of the protein, or white, phase. If the polysaccharide phase continues to increase, the structure of the phase-separated system will suddenly change, from a system in which the black phase is dispersed in the continuous white phase to

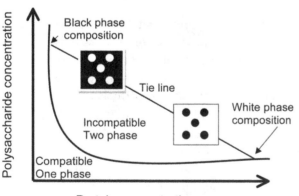

Figure 9 A typical phase diagram of polysaccharide–protein incompatibility. Certain compositions remain as one phase, but higher concentrations of either protein or polysaccharide tend to phase-separate, and the final composition of each separated phase is determined by where it falls on the tie line.

the inverse of this process. The white phase will then be dispersed in the black phase. In technical terms, this phenomenon is called phase inversion.

Another consequence of phase separation is phase concentration. This means that the concentration of, for example, the polysaccharide in its phase in the phase-separated system is greater than the nominal concentration in the formulation. Therefore, salep can be replaced only by other, more common ingredients if the phase separation potential for mixing with milk protein can be matched, along with matching the physical properties of the polysaccharide phase.

Two processes that must also be kept in mind when making such formulation considerations are those of heating, typically for pasteurization purposes, and freezing—which is, of course, essential in making ice cream. During heating, water will evaporate, thus increasing the solution concentrations of both the polysaccharide and protein from their starting formulation concentrations. In freezing, water is removed from the salep-containing phase in the form of ice, giving an even more concentrated unfrozen matrix phase within the ice cream product. Therefore, the hypothesis is that the specific stretchy textures are a result of controlled changes in the concentrations in the polysaccharide and protein phases. If the concentration of either the polysaccharide or the protein is too high, then the stretchability of the product is not optimal. Additionally, if locust bean gum is used to replace salep, stretchy textures are not possible because locust bean gum gels during the freezing process.

To test this hypothesis, we heated mixtures of milk and salep or guar gum in a rotary evaporator to drive off water and concentrate the mixture. When the approximate water content of the corresponding unfrozen matrix (in ice cream) was reached, stretchy textures were observed, confirming our hypothesis. This knowledge turned out to be the foundation to successful salep replacement in salep dondurma. In addition, it demonstrates that ice formation is not necessary to create such textures; in other words, we now can make other products with stretchy textures. In addition, it shows the two conditions that polysaccharides have to fulfill to provide salep-like stretchy textures: first, they should show the appropriate phase-separation behavior in the presence of proteins, and second, they should not gel.

Therefore, we can prepare stretchy textures on a kitchen scale, with application to sweet and savory products alike. Furthermore, using the same concepts, we can now control the texture of a wide range of dairy-based products (white sauces, cream cheeses) and soy-based products. Several of the polysaccharides mentioned may be employed as ingredients to create the stretchy textures; however, chefs and cooks can find them difficult to obtain. Alternative sources, which are more readily available, are possible, however. For instance, when mixed with proteins, mucilage gums produced from the cooking of okra provide stretchy textures in kitchen preparations. Okra polysaccharide, exuded from the okra seed pods during cooking, is known to be slimy and shows a tendency for structure formation. Also, it stabilizes foams and can form gels. Therefore, I propose a savory dish in which okra is cooked in the presence of a protein source, like soy flour. The interaction and incompatibility of the naturally exuded polysaccharide mucilage with the protein can then induce the desired stretchy behavior. Along with such an experimental approach in the kitchen, the simple addition of 0.5 gram of, for example, guar gum and ½ ounce (15 g) sugar to 3½ ounces (100 mL) of milk or soy milk is sufficient to provide stretchy textures upon either heating or freezing.

It is fascinating to realize how through a fundamental understanding of the physical world, seemingly different food products have more in common than one would ever imagine. It is now your turn to stretch your imagination.

Further Reading

Foster, Tim. 2007. "Structure Design in the Food Industry." In *Product Design and Engineering*, vol. 2, *Raw Materials, Additives, and Applications*, edited by Ulrich Bröckel, Willi Meier, and Gerhard Wagner, 617–629. Weinheim: Wiley-VCH.

seven

MOUSSAKA AS AN INTRODUCTION
TO FOOD CHEMISTRY

CHRISTOS RITZOULIS

popular culture portrays the chemist as a frantic robe-clad individual surrounded by (occasionally exploding) test tubes, beakers, and flasks filled with boiling green liquids, constantly preoccupied with the transmutation of one material into another. Is this far from the truth? Well, most of the time, the research chemist conforms pretty well to this description. Chemistry has to do with changing existing materials into new ones: iron into steel, petrochemicals into plastics, and certainly, raw into cooked food. The primeval manifestation of the research chemist is nearly synonymous with the frantic robe-clad cook surrounded by (occasionally steaming, boiling, even exploding) saucepans, frying pans, and baking sheets filled with boiling liquids, gels, and solids, and constantly preoccupied with altering the chemical composition and hence color, odor, taste, and mouthfeel of the raw materials. Cooking differs little from chemistry in terms of isolating a substance from a plant and then chemically modifying it to create a new product. To better understand this intersection between food and science, we follow a recipe on the molecular, colloidal, and macroscopic levels during the course of cooking. The term *colloidal* broadly refers to sizes in between the macroscopic level (that is, visible to the naked eye) and the molecular level, the smallest size of matter that still renders the material with all its characteristics. To host the narrative, I have chosen moussaka because it is a complex dish with a variety of ingredients and preparation stages. These multiple stages allow for a large variety of case studies, highlighting the kitchen's remarkable chemical laboratory. Experts in the kitchen, like those in the lab, manipulate ingredients on the molecular and colloidal levels, forcing them into new physicochemical entities.

Our narrative begins with the origin of the word *moussaka* itself, which, contrary to popular belief, is not Greek. At least, it does not appear to convey a Greek etymology. However, in the eastern Mediterranean and in the Near East, layering eggplant (aubergine) with chopped meat (the main element of moussaka) is at least as old as the cultivation of the eggplant. An early form of the recipe is to be found in *Kitab al-Tabikh* (*The Book of Dishes*, 1226), written by Muhammad bin al-Hasan bin Muhammad bin al-Karim al-Baghdadi. In "constructing" the dish prior to cooking, the author indicates the putting down of a base layer of half-boiled eggplant and onion. On top of this, a layer of chopped meat is added, then eggplant and onion again, and so on, until a number of layers have been formed. Dry caraway, cinnamon, coriander, cumin, ginger, pepper, and salt are dusted over each layer of chopped meat. Then saffron is added to high-quality vinegar diluted with water. This mixture is poured over the top layer of the meat and other ingredients, and then the dish is cooked. The name of the recipe is maghmuma or muqatta'a, the latter label bearing some resemblance to the word *moussaka*.

The modern recipe is composed of raw materials originating from places as diverse as South America and Southeast Asia. Most spices, of course, come from Southeast Asia. Eggplant was known in the southeastern Mediterranean and Middle East by late Roman times, where the plant had been introduced via the spice road that joined Indonesia and India to the Red Sea and hence the Mediterranean. Eggplant was well known to both Byzantines and Arabs, and it has always been considered a luxury upper-class food. Potatoes, a later ingredient in moussaka, originated in the Andes Mountains on the western coast of South America and had come to the eastern Mediterranean by the seventeenth century. By the eighteenth century, potatoes were not considered exotic to Ottoman Constantinople (now Istanbul). In the southern, more isolated parts of Greece, potatoes were introduced more recently, apparently during the nineteenth century. I believe potatoes found their way into moussaka at a much later stage.

Perhaps the most recent addition to moussaka is the béchamel (white) sauce. In the early twentieth century, Nikolaos Tselementes, a French-influenced chef, wrote a series of cookbooks so successful that in modern-day Greek, the word *Tselementes* is practically synonymous with *cookbook*. The master chef was among the first to advise the use of béchamel sauce as the final topping on the eggplant–chopped meat–potato dish, thus giving us moussaka as we know it today.

Gastronomy is dynamic: no recipe has been or ever will be carved in stone. The way we cook moussaka today will certainly change in the next few hundred years. Retaining some mental placeholders to allow for additions in the years to come, this is the dish as we currently know it.

Now, let us reexamine the process from the start, a bit more diligently, starting from the chopping of the ingredients.

As the onions are chopped, their cells are torn apart and the enzymes previously isolated in restricted parts of the cells now get intermingled with other substances. These enzymes readily turn some sulfur-containing compounds into volatile mol-

MOUSSAKA

4½ pounds (2 kg) eggplant, preferably large
2¼ pounds (1 kg) potatoes, preferably large
2¼ pounds (1 kg) tomatoes
5¼ ounces (155 mL) olive oil
2 large onions
2¼ pounds (1 kg) chopped veal
Salt and pepper

Béchamel Sauce

¾ ounce (21 g) butter
1 ounce (28 g) all-purpose (wheat) flour
1 pint (500 mL) milk

Wash, peel, and cut the eggplant and the potatoes into long, thin slices. Salt them *and* immerse them in salted water (about ¾ ounce [21 g] salt per quart of water). In a food processor, grind the tomatoes, separating some 3½ ounces (100 g) of core and seed (that is, without the watery liquid).

Prepare the béchamel in a saucepan. Begin by melting the butter at a low temperature and preparing a light roux. Slowly blend the flour into the butter, stirring so that they froth together for 2 minutes and taking care not to let the mixture brown. Remove the béchamel from the heat and put it aside so that the bubbling stops.

Meanwhile, heat the milk to 180°F (82°C). Add the milk to the roux all at once. Beat the mixture vigorously with a wire whisk to blend the liquid and roux, making sure to gather all the grainy bits sticking to the pan. Set the sauce over moderately high heat and stir with a whisk until it comes to a boil. Boil for 1 minute while stirring, remove from the heat, and add salt and pepper to taste.

Heat some olive oil in a frying pan. Stir-fry the onions. When they become golden brown, add the 3½ ounces of reserved tomato core and seeds. When these dry up and start to fry, stir in the chopped veal. Add salt to taste and shallow-fry until the color changes from red to brown. Add pepper and the rest of the ground tomato (including the liquid) and leave simmering on medium-low heat for half an hour, until most of the water is absorbed. Add olive oil in another pan and fry the eggplant and potatoes. Remove the fried eggplant and potatoes to a stainless steel sieve for draining.

Spread 2 tablespoonfuls of olive oil on the bottom of a large baking pan and lay in the potatoes slices to form a solid layer. Spread half the chopped meat across the potatoes. Then place a layer of eggplant on top of the chopped meat. Add the rest of the meat on top. Cover with another layer of eggplant. Finally, pour the béchamel in a thick layer over the eggplant. Bake at 350°F (180°C) for 30 to 40 minutes, or until the surface of the sauce is dark brown. Remove from the oven, allow 10 minutes for the béchamel to set, and then cut into portions and serve.

ecules. The flying onion molecules seek and dissolve into any aqueous solution around—yes—including the tears in our eyes. Upon contact with the eyes, the onion molecules stimulate sensory neurons, creating a painful burning sensation. We reflexively rub our eyes with our fingers. Alas, our fingers being saturated with sulfur compounds from the sliced onion only makes things worse. Our reflexes do not stand idle, however: tear production immediately commences, and the trouble-making molecules get diluted and washed away.

Unlike the onion, chopping up the eggplant and the potato will not make us cry. This action will, however, cause the vegetables to turn black because chopping destroys cell walls. In defense, the cells release specialized enzymes that readily look for phenolic compounds, which abound in these vegetables. The enzymes will turn these phenolic agents into dark and sometimes bitter compounds, which are an antimicrobial defense line of the injured cells. Still, they spoil the look of our food. Cut up an apple or a banana, and leave it out: enzyme activity will soon turn it dark. The best way to avoid this is to denature, or inactivate, such enzymes (that is, destroy their shape and, essentially, their functionality). Denaturing is easily achieved by heating (this is why we blanch vegetables) or by exposing the enzyme-rich food to a solution of high salt concentration (brine). This is one of the reasons why salty water will preserve the color and taste of eggplant and potatoes.

Of course, a good deal of other enzyme-denaturing techniques are used in a range of foods. Take pickled foods, for example: enzymes, being basically proteins, are sensitive to changes in pH; exposing food to acidic solutions, such as lemon juice or vinegar, will denature and render these oxidizing enzymes inactive, despite the rupture of the cell walls, thus preserving the color of the food.

Now, let us heat up some olive oil and stir in the onions: chemical reactions that take place within the onions lead to darkening, the development of a rough texture, and the formation of small volatile molecules that give the characteristic flavor of do-pyaza (a Punjabi okra dish) or stifado (a Greek meat stew). How is this new color, flavor, and texture created? The basis of these changes from raw to cooked food is a series of very complex chemical reactions between the amino-acid components of the proteins and the individual sugars of the carbohydrates, collectively known as the Maillard reaction (for a detailed account of these reactions, see chapter 13). These reactions can follow a wide range of paths, leading to different final products, depending on the composition of the sugars and proteins, the pH, the temperature, and of course, the duration of cooking. As a simple example, among the end products of the reactions are volatile molecules that can be perceived by our noses as the familiar pleasant smell of food being cooked.

Maillard and caramelization (degradation reactions involving only sugars) reactions lead also to the development of color: one may say that the dark brown color of fried onions, the brown-black color of roasted meat, and the dark brown color of roasted béchamel are, to a large extent, due to products of such reactions. The molecules in the food yield color depending on their chemical structure. As white light shines onto them, they absorb specific colors, while at the same time they reflect the other colors. As new molecules are formed during cooking, new colors are absorbed by the food. As more and more colors are absorbed during cooking, less are reflected, leading to a loss of color during the later stages of cooking. This is how most foods turn darker as cooking proceeds. Food texture (that is, development of crust) and, occasionally, development of new flavors are also related to Maillard and caramelization reactions.

So far, we have visited the world of molecules one-billionth of a meter in size. We have addressed things such as color and smell, all related to the world of molecules. Other amazing events are bound to unfold, but to witness these events, we must zoom out to the scale of microns (one-millionth of a meter). A brave new world opens up, composed of huge agglomerates of molecules, oil droplets, and tiny bubbles. And there is an entity out there that has it all: béchamel.

As we have shown, béchamel sauce is made by mixing roux and milk. Roux is a mixture of fat (from butter) and flour, which contains particles with long chains of molecules (originating from gluten proteins and starch carbohydrates). Milk is a dispersion of fat droplets in a solution of sugars, proteins, and other minor constituents. Mixing of the two results in a complicated colloidal system containing all these different particles and molecules. In addition, whipping can incorporate some air droplets. So, béchamel could also be called a foam.

Cooked béchamel, basically long chain molecules (and smaller molecules) in solution together with a number of fat droplets, is elastic: if one applies a gentle pressure to the sauce, say, with the back of a flat spoon, it will deform, only to spring back once the spoon is withdrawn. Resistance to the movement of the spoon will increase as the deformation increases. That is called elastic behavior, which is characteristic of solids. So, is béchamel solid? Well, if one applies a greater pressure by hanging a set béchamel in the air from one side, like a carpet, allowing it to hang loose, the béchamel's own weight will cause it to elongate. If we put this poor béchamel back on the table, it will not spring back to its original length. An amount of the energy applied to the béchamel has been used to break bonds that held the former tight structure together. This is a typical behavior of fluids, so béchamel can be said to be both solid and liquid. To a large extent, it is the delicate interplay between the solid and liquid character that gives every food its particular texture and mouthfeel.

As we finish what we started, from a historical narrative of a dish followed by a Tselementes cookbook recipe, we can summarize our trip in this way: we have zoomed out from the molecular level (structure, light–matter interactions) to the colloidal level (long chains of sugar-type molecules, fat droplets, air bubbles) and in the end to what we call the macroscopic world (shape, color, smell, taste, texture). The examples in this chapter serve to demonstrate that cooking, along perhaps with metallurgy, are the earliest true forms of chemistry and that no culinary skill is irrelevant to a laboratory activity. In fact, chemistry, like its cousin medicine, spent most of its lifetime as an empirical art rather than a science. And, indeed, chemistry has been successfully explored and exploited by cooks, metallurgists, dye makers, and alchemists. The good news is that a cook can lay claim to chemistry as much as an alchemist, a dyer, or a metallurgist ever could.

Further Reading

Belitz, H.-D., W. Grosch, and P. Schieberle. 2004. *Food Chemistry*. Translated by M. M. Burghagen. 3rd ed. Berlin: Springer.

Coultate, Tom P. 2009. *Food: The Chemistry of Its Components*. 5th ed. Cambridge: Royal Society of Chemistry.

Dickinson, Eric. 1994. *An Introduction to Food Colloids*. Oxford: Oxford University Press.

Shallenberger, R. S. 1993. *Taste Chemistry*. Glasgow: Blackie Academic & Professional.

eight

THE STICKY SCIENCE OF MALAYSIAN DODOL

ALIAS A. KARIM AND RAJEEV BHAT

malaysia is a melting pot of ethnic backgrounds, which is reflected in the diverse food culture. Its cuisine stems from the unique combination of Malay, Chinese, Indian, and Thai influences. Eating out is a gastronomic adventure. One can find an immense variety of dishes ranging from spicy Malaysian and Indian to even Portuguese! Apart from the ubiquitous savory and spicy dishes, Malaysian cuisine has an alluring array of traditional sweet delicacies. In this chapter, we describe one of them—dodol, a sticky, sweet rice-based dessert—and the basic science involved in its preparation.

Dodol

Dodol is a popular delicacy in Malaysia and other Asian countries, such as Indonesia, Thailand, and the Philippines. It is eaten especially during religious Muslim festivals. In the ancient Malay community, dodol was used as a gift to promote unity and strengthen relationships.

Dodol is a starch-based dark brown product with a soft, chewy, and sticky texture. It is packaged into different shapes and can be kept at room temperature (77–86°F [25–30°C]) for at least four to six weeks. It has a moisture content of about 45 to 50 percent. The main ingredients of dodol are coconut milk, glutinous rice flour, regular rice flour, palm sugar, and pandan leaves (from the pandan plant). Typically, the glutinous rice and regular rice flours are mixed at a 6:1 ratio.

The process of making dodol is simple but time consuming, as it takes up to 6 hours to prepare. A typical recipe of 4 pounds and 6½ ounces (2 kg) glutinous rice and regular rice flours requires approximately 2 quarts (2 L) of thick coconut milk. This amount of coconut milk can be obtained from 10 to 12 mature coconuts (or simply from canned coconut milk). Usually, a portion of the coconut milk is used in combination with the flour to prepare a smooth slurry. The slurry is then strained into a large steel wok. Meanwhile, palm sugar is boiled in approximately 1½ quarts (1.5 L) of water together with the pandan leaves until the sugar is dissolved completely. The sugar solution is strained into the flour mixture and mixed well. The mixture is stirred over low heat until the volume of the liquid has been reduced by half. Thick coconut milk is then added. The mixture is stirred continuously until a very thick, dark, and shiny mass has been formed (figure 10).

Let us now examine and decipher the role of each ingredient in making dodol so as to better understand the physical changes that occur over its long cooking process.

Rice Flour

The main constituent in rice flour is starch. Minor constituents are proteins and lipids. Starch consists mainly of two types of long-chain molecules or polymers: amylose and amylopectin. Amylose is essentially a linear polymer, whereas amylopectin is a branched molecule. Amylopectin has a larger mass than amylose. These structural differences contribute to the significant differences in their properties and functionality. The amount of amylose varies from between 15 to 30 percent of the total starch; the other 85 to 70 percent is amylopectin. Under the microscope, we can see discrete particles between 2 and 10 microns (1 micron equals 0.00004 inch [0.001 mm]) in length, often clustered together. We could imagine that in the old days, different kinds of starch may have been tried, finally leading to the use of rice flour as the best starch to provide the desired smooth and elastic textural attributes. The texture of dodol would have been very different if corn, wheat, tapioca, or potato had been the starch of choice.

Figure 10 Different stages in the cooking of dodol: (*a*) the rice flour, coconut milk, and other ingredients become a thin slurry during the early phase of cooking; (*b*) the mixture increases in viscosity from starch gelatinization; (*c*) as cooking progresses, the mixture turns into a thick brown mass from a complex Maillard reaction and caramelization; (*d*) after a few hours, the mixture is a highly viscous, dark brown mass.

The traditional dodol recipe contains a mixture of glutinous rice flour and regular rice flour at a 6:1 ratio. What is the rationale for using such a flour mixture? Since ancient times, many types of traditional foods from the region have been made from mixtures of different starches. For example, in the production of rice vermicelli, at least two other starches, usually corn and sago (from the pith of sago palm stems), are used together with rice flour. Likewise, fish crackers are often made from a mixture of tapioca and sago starches. The starch mixtures convey the desired textural and sensory attributes that cannot be created by the individual starches alone. In the dodol recipe, glutinous rice flour and regular rice flour each offer unique and distinct characteristics to the final product.

The major component in glutinous rice flour is waxy starch. As the name implies, waxy starch produces a long and stringy paste upon cooking. Therefore, the role of glutinous rice flour in dodol is to impart its characteristic elastic and chewy texture. Compared with waxy starch, nonwaxy starch in regular rice flour usually exhibits a less stringy paste upon cooking. Hence, the role of regular rice flour is to modulate the overall texture of the product so that it is not too elastic/chewy or too firm.

During the cooking process, it is interesting to observe the transition of the slurry from thin and white (figure 10a), to a slightly viscous brown paste (figure 10c), and, finally (after 5 to 6 hours), to a very thick and viscous dark brown mass (figure 10d).

The increase in thickness during the initial stage of cooking (figure 10b) is mainly due to the starch in the rice flour. A series of changes, together known as gelatinization, occur in the starch granules when they are heated in the presence of excess water. The term *gelatinization* is used to describe a series of events, such as the swelling of the granules and an increase in viscosity. Figure 11 shows several stages in the gelatinization of starch (for potato starch in this example). Upon heating, the granules begin to swell like balloons; a granule can swell to about five times its original size. At this point, the starch paste becomes thicker. Some granules appear deformed and folded (figure 11d). Glutinous rice is a waxy variety of cereal starch, and such starches generally begin to swell at lower temperatures than other similar starches.

Immediately after cooking, dodol exhibits as a very viscous and sticky mass. However, upon cooling, the mass achieves a firmer gel-like texture that basically can be cut into freestanding shapes. What are the underlying physical and chemical changes during the transformation of hot viscous mass into a firm gel? Actually, during cooling, the starch molecules reassociate and form an ordered crystalline structure. This is called retrogradation. When this process occurs in bread, we say that the bread has gone stale. The ratio of glutinous rice flour to regular rice flour in a dodol recipe has to be adjusted in such a way that the final outcome of retrogradation results in a desirably soft and chewy texture. The addition of too much regular rice flour results in a firmer and less chewy texture, whereas too much glutinous rice flour makes dodol too soft, stringy, and sticky.

Palm Sugar

The name indicates sugar's source—the palm tree. Palm sugar is commonly processed from the sugary sap of the coconut or sago palm.

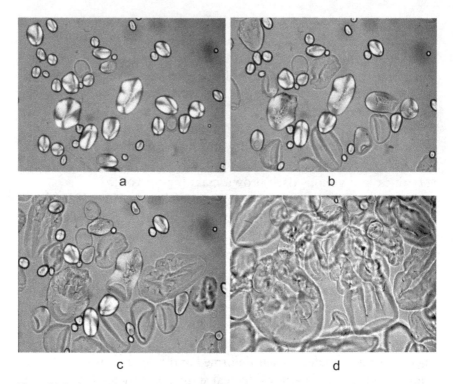

Figure 11 During cooking in excess water, starch granules undergo several physical changes that can be viewed using a polarized light microscope: (*a*) starch granules at the beginning of heating; (*b*) granules begin to swell; (*c*) as heating progresses, starch granules attain maximum swelling and begin to burst; (*d*) in the presence of shear (stirring), most granules are broken. (Photographs courtesy of Professor Mike Gidley)

In Malaysia, palm sugar is known as *gula melaka*. Basically, palm sugar imparts not only sweetness to dodol but its characteristic flavor and dark brown appearance. Palm sugar contains common sugars, such as glucose, fructose, and sucrose—or table sugar. These sugars, especially glucose and fructose, play an important role in flavor and color development during the first few hours of the cooking process via a complex series of chemical reactions, collectively known as the Maillard reaction (chapter 13). These reactions are prevalent in almost all heat-processed foods. Apart from inducing browning, they have a significant impact on flavor. For the Maillard reaction to take place, sugars are required as well as amino acid groups, which are provided by proteins. The proteins stem from the rice flour and coconut milk for the most part. It should be noted that in addition to the Maillard reaction, sugar caramelization can occur in dodol, although this type of reaction is usually associated with dry heating or toasting in the presence of a high sugar concentration. The final outcome of both Maillard and caramelization reactions is myriad compounds that together contribute to the characteristic dark brown color and flavor of dodol, which can be described as cooked, roasted, toasted, and even caramelized. Another point worth mentioning is that palm sugar itself contains both Maillard reactions and caramelization products because during the production process, the sugar mass is heated at high temperatures.

Coconut Milk

This is one of the most common ingredients in Southeast Asian dishes and also one of the key ingredients in dodol. The milk is extracted from mature coconut meat. The average composition of freshly extracted and undiluted coconut milk is 47 to 56 percent water, 27 to 40 percent fat, and 3 to 4 percent protein. In dodol, the fat component of the coconut milk plays a couple of functional roles:

1. The formation of a fatty acid–amylose complex reduces the retrogradation of starch, thus modulating the hardness and elasticity of dodol.
2. The coconut fat imparts dodol's shiny appearance and overall flavor and mouthfeel.

Pandan Leaves

Leaves from the pandan plant (*Pandanus amaryllifolius*) are often used in many types of traditional savory and sweet Southeast Asian dishes, such as puddings, to lend a characteristically fragrant aroma and flavor reminiscent of freshly cooked jasmine rice and to provide an attractive green color. For example, in making jelly agar, pandan leaves are usually boiled together with agar and sugar to extract the flavor and green color. The pandan plant grows in tropical areas, including the Pacific islands, South Asia, and Southeast Asia. In the dodol recipe, pandan leaves are added as an optional ingredient mainly for the flavor rather than the color; this is because the green color of pandan extract is effectively masked by the dark brown shade of the final product. The addition of pandan leaves yields various compounds, such as essential oils, carotenoids, tocopherols and tocotrienols, quercetin, and alkaloids. Together with the numerous products created from the Maillard reaction, complex interactions among the various compounds are inevitable. At high temperatures and long cooking times, many compounds are produced that reflect the characteristic flavor of dodol.

A Final Reflection

We have provided a brief account of the chemistry and physics behind the preparation of dodol, one of the traditional sweet delicacies of Malaysia. It is evident that almost every attribute (texture, color, flavor) of dodol is the consequence of a complex series of reactions from a short list of common ingredients. In dodol, starch in particular plays a central role in controlling the textural attributes, whereas coconut milk and palm sugar provide the characteristic flavors and color. From the chef's perspective, an understanding of the science behind the process allows for informed manipulation of the recipe.

nine

THE PERFECT COOKIE DOUGH

AKI KAMOZAWA AND H. ALEXANDER TALBOT

our web site, ideas in food (www.ideasinfood
.com), is all about improvisation and experimentation in the kitchen.
We believe in the marriage of science and creativity as it pertains to
cooking and to all aspects of life. People who visit our Web site are
sometimes baffled by the wide-ranging topics, from the origins of
ramps and the use of hydrocolloids, to serving a beautiful piece of
meat or fish as simply as possible, to making kimchi cracklings and
our signature combination of caviar and ice cream. People sometimes
have trouble assimilating these seemingly schizophrenic combinations. The unifying thread is that food should always be tasty and
delicious—by whatever method strikes us on a given day.

As we travel the road from cook to chef, we pass through a series
of revelations. First, we learn how to cook by gaining a basic understanding of the majority of the different cooking techniques: roasting,
sautéing, braising, baking, and frying. Then we learn to substitute
ingredients to create new and interesting dishes extrapolated from
classic recipes. Then, consciously or subconsciously, a lightbulb flick-
ers on, and we realize that cooking is all about ratios (for a detailed
discussion of the rationale and utility of ratios, see chapter 33). Once
we figure out the particular ratios involved in any individual preparation, we are free to create endless variations on the theme. Cooks

come to this realization more quickly in the pastry kitchen because recipes are actually taught in ratios there. Eventually, anyone who spends time learning to hone his or her craft in a professional kitchen makes that critical leap of understanding: once you know how to lay tile, you can use any kind of tile you want, as long as your foundation is sound.

A smart cook—one we would be happy to have in our kitchens—pays attention to the world around him or her. Professional kitchens move at a breakneck pace, and the best cooks are masters of efficiency. The way to become efficient is to use all five senses and to understand what happens to food as it is being cooked, such as knowing that when food is gently dropped into a fryer, there is a loud bubbling sound and roiling motion as the hot fat meets the food. This is because there is always water on or near the surface of the ingredients that evaporates upon contact with the hot fat. As the food cooks, the noise slowly dissipates and, when things slow to a gentle bubble, most of that surface water has disappeared, resulting in a crisp, dry coating as the oil is drained off. The sound of sizzling in a pan meant to simmer will alert you to a sudden drought. The touch of a steak will tell an experienced cook if it is ready to be plated. A good nose will detect the smell of browning butter before vegetables start to burn in the pan. A sharp eye will gauge the perfect jiggle in the custard and pull it from the oven before it becomes a scrambled mess. The knowledge is instinctual, seemingly absorbed from the atmosphere in the kitchen. In reality, it is derived from unconsciously observing, actively paying attention, and learning from both success and failure. This hard-won store of information is a chef's best tool in the kitchen.

In 2008, David Leite wrote about the search for the perfect chocolate chip cookie. His *New York Times* article created a firestorm online because the technique was so different from any that came before it. Leite presented what he considered to be the quintessential chocolate chip cookie, a recipe adapted from Jacques Torres, the master pastry chef and chocolatier. The secret? Letting the dough rest in the refrigerator for 24 to 36 hours. This resting period allows the dough to fully hydrate, which in turn causes the cookies to have a fuller, more complex flavor. The other secret to his cookie recipe is size: a larger cookie allows for a greater range of textures and a more pleasing final product.

We were immediately caught by the idea of hydration. We often make cookie dough in advance, more for efficiency's sake than for flavor. We prefer the texture of cookies made from cold dough to that of cookies made straight from the mixer. In retrospect, we realized that the flavor of the freshly mixed and baked cookies was flatter and sweeter than the flavor of cookies made with cold dough, but we had never really given the concept of temperature much thought. Now, the idea that hydration could amplify the flavor of cookie dough ignited our imaginations, and we immediately looked to the vacuum sealer.

The vacuum sealer is primarily used to preserve freshness in food. Here, we use a specific model of vacuum sealer in which the plastic bag containing the product to be vacuum sealed is placed inside a vacuum chamber. Outside air is sucked from the chamber and from the filled storage bag sitting within the chamber, creating a low-pressure environment. This environment allows cell walls to expand without atmospheric constraints. Once the air is removed from the chamber, the bag is sealed. Then the air is returned to the chamber and the bag collapses around its contents, uniformly squeezing the contents flat. This oxygen-reduced environment allows food to stay fresh for a longer period of time. Many modern vacuum sealers permit the user to program different levels of atmospheric pressure so that the process can be tailored to the ingredient. You can achieve similar results with vacuum cells or vacuum ovens, which replicate this process without sealing ingredients in a bag. However, be aware that this will not work with vacuum sealers in which the bag containing the cookie dough remains external to the vacuum and sealing device—in other words, the dough is not subjected to low pressure.

A nontraditional use for the vacuum sealer creates nearly instantaneous hydration in porous foods, where capillary forces can help distribute the water. Once the air is removed from the dough, the moisture is evenly distributed throughout the preparation. We discovered this by chance when we vacuum sealed some pasta dough for storage. We were struck by the change in the dough; its color bloomed and texture softened as a result of its time in the vacuum chamber. It was as though the normal resting period had taken place in the blink of an eye.

It was one small step to apply this technique to cookie dough. This was necessary because who really wants to wait two days for chocolate chip cookies? When we make dough at home, immediate gratification is more of an issue. We whipped up our favorite recipe and vacuum sealed half of the dough and wrapped the other half in plastic wrap.

The color of the vacuum-sealed dough immediately became richer and more vibrant—and the texture changed, becoming softer and more elastic (figures 12 and 13). Then we made the ultimate sacrifice for the sake of science and placed both sets of dough in the refrigerator.

Figure 12 Two doughs: (*left*) traditional and (*right*) vacuum sealed.

Figure 13 Two scoops of dough: (*left*) traditional and (*right*) vacuum sealed.

Thirty-eight hours later, we made another batch of dough, again vacuum sealing half and wrapping the other half in plastic wrap. Then we put both sets in the refrigerator for about an hour to chill because we think cold dough makes better cookies. Then, finally, we baked cookies from both sets of both batches of cookie dough. We had two baking sheets, one for each batch, with the plastic-wrapped cookie dough at one end and the vacuum-sealed dough at the other.

We made our normal-size cookies because we were focusing on the hydration aspect of the technique, and the variations were obvious as soon as we pulled them from the oven. The vacuum-sealed cookie dough on both sheets was noticeably darker and slightly shinier than the plastic-wrapped dough. It also seemed to have spread out a tiny bit more. All of the cookies were slightly soft in their centers and the plastic-wrapped cookies were tender and cakey throughout. The vacuum-sealed doughs produced slightly chewier cookies with edges that were more

CHOCOLATE CHIP COOKIES

8 ounces (225 g) unsalted room-temperature butter
4 g baking soda
4.5 g fine sea salt
7½ ounces (210 g) golden brown sugar
4 ounces (110 g) dark brown sugar
2 large room-temperature eggs*
5 g vanilla extract
12 ounces (340 g) all-purpose flour
8¾ ounces (250 g) semisweet chocolate chips[†]

Cream the butter, baking soda, and sea salt using either a hand mixer or a stand mixer with the paddle attachment until thoroughly blended. Add the golden and dark brown sugars, 3–4 ounces at a time, and blend until light and fluffy. Pour in the vanilla extract. Add eggs, one at a time, making sure the first egg is thoroughly absorbed before adding the second egg. Once the second egg has been fully absorbed, stop the mixer and add all the flour to the bowl. Mix on low until the dough just comes together. Fold the chips into the dough using a rubber spatula. Vacuum seal the dough[‡] and use immediately or chill until needed.

Preheat oven to 375°F (190°C).

Line two or three baking sheets with parchment paper or silicone mats. Using approximately 2 tablespoons per cookie, shape the dough into balls either by hand or by using an ice cream scoop, then flatten them. Place the flattened dough balls on the baking sheets, leaving 2 inches (5 cm) between, which will give them room to spread in the oven. Bake for 5 to 8 minutes (7 to 10 minutes if using chilled dough), rotating the baking sheets once, until the cookies are golden brown and no longer appear wet in the center. Let cool on the sheets for at least 5 minutes before removing them to a rack or before eating them.

* If you forget to pull them out in advance, you can hold them in a bowl of warm water for a few minutes to bring them to room temperature.

[†] Ghirardelli is our house brand.

[‡] For our experiment, we used a Berkel 250 vacuum sealer and ran the chamber for 2 minutes.

crisp and caramelized. The difference in texture occurred because the starch in the flour became fully hydrated.

Across the board, the chocolate was gooey and still warm, and there were buttery caramel notes to the cookies themselves. The vacuum-sealed cookies seemed to have an added richness, with the butter flavor more pronounced, on the palate. There seemed to be less of a difference between the two plastic-wrapped batches of dough and the two vacuum-sealed batches of dough, with the most striking deviations occurring in each vacuum-sealed batch. In our experiment, it was the hydration that occurred in the vacuum chamber that made the real difference in the finished product. Compared with that distinction, the variance between the two types of dough with only a rest period as a factor was minimal. There are lots of ways to add depth to your cookies. You could try different chocolates, better butter, and higher-quality sugars, or you could just vacuum seal the dough and enjoy the benefits of technology.

Further Reading

Leite, David. 2008. "Perfection? Hint, It Is Warm and Has a Secret." *New York Times*, July 9.
"Vacuum Sealed Cookie Dough." Available at http://blog.ideasinfood.com /ideas_in_food/2008/07/vacuum-sealed-cookie-dough.html.

ten

TO BLOOM OR NOT TO BLOOM?

AMELIA FRAZIER AND RICHARD HARTEL

in the kitchen, we often take for granted that seemingly simple culinary projects will go off without a hitch. Take, for example, chocolate chip cookies. For most of us, making ordinary chocolate chip cookies is not much of a challenge. Just follow the simple recipe on the bag of chocolate chips and even the novice cook can make delicious cookies. But there's more to these cookies than just scooping and baking. Upon closer examination, some fairly complex physical chemistry occurs within a cookie (some of it shown in chapter 9). But why is it that the chocolate chips in cookies do not "bloom" after the cookies have been baked and cooled? To bloom or not to bloom—that is the question.

What is bloom, exactly? Think of a chocolate bar that has melted in a hot car on a warm summer day. When it cools and resolidifies, grayish-white streaks appear within a day or two. This is one form of chocolate bloom. It may look like mold, but it is simply cocoa butter that has not crystallized properly. Bloomed chocolate will not hurt you, but it looks unappetizing and loses that perfect flavor release you expect when well-tempered chocolate melts in your mouth. When you bake cookies, the chocolate chips melt completely, just like the chocolate bar left in a hot car. Why does one bloom but not the other?

When chocolates are molded, the molten liquid must be tempered prior to depositing. Tempering is the process of manipulating the final arrangement of cocoa butter fat crystals through a controlled process of cooling and rewarming. Cocoa butter can crystallize in multiple crystalline forms called polymorphs, only one of which gives chocolate its desired properties. Good tempering produces lots of very small cocoa butter crystals in the appropriate crystalline polymorph. When done properly, well-tempered chocolate releases easily from a chocolate mold, gives a firm snap when broken, and has a glossy sheen. If solidified incorrectly, like the melted and resolidified chocolate in the car, bloom forms on the surface and the chocolate has an undesirable soft, plastic-like texture. In this case, an unstable polymorph crystallizes first, followed by the transformation into a more stable polymorph over a period of hours to days. The polymorphic transition results in the contraction of the cocoa butter crystal matrix, which in turn pushes cocoa solids and sugar particles to the surface of the chocolate. It is these particles that are responsible for the whitish appearance indicative of untempered chocolate bloom, as seen in figure 14, which shows a bloomed disk of chocolate that was not properly crystallized.

Baking a chocolate chip cookie is very much like leaving your chocolate bar in a hot car—almost. When baked, the chocolate chips in the cookie are subjected to high temperatures that cause them to melt completely, destroying the desirable cocoa butter crystals. As the cookie cools, the chocolate chips resolidify. Just like the chocolate left in the car, the cocoa butter crystallizes improperly. Despite this, bloom on chocolate chips baked in cookies is rarely observed. The kitchen scientist may ask, What is it about a chocolate chip cookie that inhibits chocolate bloom on untempered chocolate chips?

Figure 14 Characteristic chocolate bloom can be seen on untempered chocolate.

Here is an experiment to try at home. Bake a few loose chocolate chips (in a paper muffin cup) on the same baking sheet next to some chocolate chip cookies. After they have cooled, observe these stray chips over a couple of days and compare them with the chocolate chips in the cookies (be sure to set a few cookies aside for this experiment). You should find that the loose chocolate chips bloom within a day or so, whereas the chocolate chips in the cookies remain unbloomed for the life of the cookie. This suggests that there may be something about the cookie itself that is preventing bloom.

A while back, a cookie manufacturer approached us with the unusual problem of bloom in chocolate chip cookies. The baker was making low-fat organic cookies with palm shortening and found that the chips in the cookies were blooming shortly after baking. The unsightly splotches on the chips displeased her customers, and not surprisingly, her business suffered because of it. With our assistance, the baker eventually remedied the situation by increasing slightly the amount of fat in the cookies, which suggested that fat content in the cookie dough could be crucial to the inhibition of chocolate chip bloom.

This led us to the hypothesis that fat migration from the cookie dough into the chocolate chip was responsible, at least in part, for preventing bloom. To test this hypothesis, we baked cookies with different amounts and types of fat, including butter, palm oil, olive oil, and vegetable shortening. We discovered that all four fats inhibited bloom in chocolate chips provided the fat content was high enough in the cookie dough. Below this fat content, chocolate chips in the cookies bloomed, just as we saw with the commercial low-fat organic cookies. The amount of fat needed to prevent bloom varied slightly with the type of oil, but they all showed the same behavior. If fat content was above a critical level, no bloom was observed. The chocolate chips in the cookie made with 14 percent palm oil (figure 15a) showed evidence of bloom, whereas those in the cookie made with 20 percent palm oil (figure 15b) were still a nice (in fact, brown) color.

To further support our fat migration hypothesis, we observed that chocolate chips protruding unevenly out of a cookie would bloom, whereas a chip in the same cookie that was surrounded sufficiently by dough would not bloom. That is, if a chocolate chip was sticking out of the cookie, the point of the chip not in contact with the dough would bloom, whereas the base of the chip would not. Clearly, we surmised, the chocolate chip must be in contact with the cookie dough in order

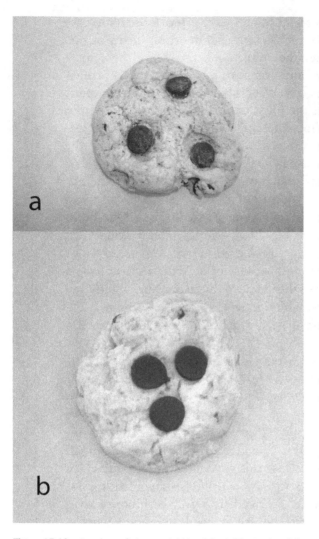

a

b

Figure 15 After ten days of storage, (a) bloom is visible on chocolate chips baked in a cookie made with 14 percent palm oil, whereas (b) bloom does not occur with 20 percent palm oil.

to receive protection from bloom. As can be seen in figure 16, the upper-right chocolate chip is not fully situated within the cookie; therefore, bloom occurs on the top, whereas the chip to the left, which is fully embedded in the cookie, shows no bloom.

One point we should mention is that during baking, fat not only migrates from the cookie dough to the chocolate chip but also from the chocolate chip to the dough. This occurs because there are two different fats present—cocoa butter in the chocolate chips and the fat in the cookie dough. When two different fats are present in dough, one fat migrates into the other until equilibrium is reached. In one experiment, we baked chocolate chip cookies using cocoa butter as the fat in the dough. Unlike the cookies we baked with butter, palm oil, olive oil, and shortening, all the chocolate chips in the cookies made with cocoa butter bloomed, no matter how much cocoa butter we added to the dough. Since the fat in the cookie dough was the same type as the fat in the chocolate chip, there was no driving force for the fats to migrate toward each other. It appears that fat migration is crucial for bloom inhibition in the chips of chocolate chip cookies.

To determine whether the fat migrating into the chocolate chip from the dough was truly controlling bloom, we experimented with mixing just the fats from the cookie dough with the melted chocolate. We found that by adding enough melted fat to the chocolate, we could inhibit bloom upon solidification. Conversely, bloom formed as expected when the added fat levels were not high enough, similar to the effect we observed with low-fat cookie dough and chocolate chip bloom. From these observations and other fat analysis methods, we determined it was the fat migrating into the chocolate chip that was influencing chocolate bloom.

Figure 16 In order to inhibit bloom, chocolate chips must have adequate contact with the cookie dough. The chocolate chip on the upper left side of the cookie was pushed sufficiently into the dough, and no bloom is visible. The chocolate chip on the upper right side of the cookie protruded unevenly from the dough, and bloom formed extensively on the surface of the chip.

From this evidence, we conclude that fat migrates into the chocolate chip during baking, preventing bloom formation. The interaction of the fat from the dough and the cocoa butter in the chocolate chips must be at least partially responsible for inhibiting bloom.

Of course, fat is not the only ingredient in cookie dough. Other components of the cookie dough, such as the flour, water, and protein, as well as environmental conditions, such as room temperature and humidity, may also influence bloom formation. In order to observe the specific effect of fat content on chocolate chip bloom, we needed to create a model system that would mimic a chocolate chip cookie without the influence of the flour, egg, and water—the other key ingredients in the dough. We created a model system consisting of washed sea sand and fat. The sand was creamed in a mixer with the fat, similar to how sugar and fat are mixed in the initial steps of making cookie dough. We scooped the sand–fat mixture into muffin cups and pushed the chocolate chips into the surface. Our "sand cookies" were baked and stored in the same manner as the regular cookies. As figure 17 shows, bloom on the chocolate chips was observed when the fat content was less than about 9 percent. Above this value, the chocolate chips retained their nice brown color.

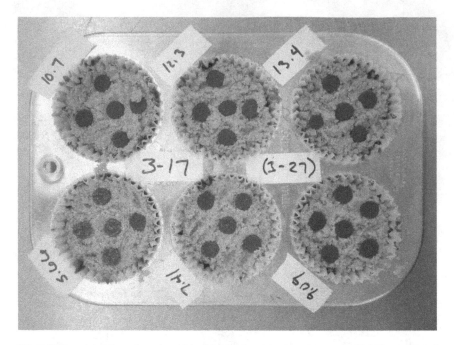

Figure 17 A model system of sand and fat mixtures: chocolate chips were pushed into the surface of sand and fat mixtures and these mixtures were baked in the same manner as the cookies. As the fat content of the sand was increased, the amount of bloom formation on the chocolate chips decreased.

As with regular chocolate chip cookies, we found that chocolate bloom was inhibited in the sand cookies at higher fat contents. However, the amount of fat required to inhibit bloom in the sand cookies was much less. This could mean that in addition to fat content, some other component of the cookie, or environmental factors, influences bloom formation. Or alternatively, our model system does not perfectly mimic a chocolate chip cookie. Perhaps fat is more easily able to migrate through sand than through cookie dough, and as a result, a chocolate chip baked in sand receives more of the migrating fat during the baking process.

In addition to observing chocolate bloom formation in chocolate chip cookies with varying fat contents, we examined chocolate chip cookies with different water and egg contents. We also looked at cookies cooled at different rates and cookies stored at both high and low air-moisture (humidity) levels. Unfortunately, our data showed no pattern that suggested a relationship between these other variables and bloom inhibition. This does not necessarily mean that these other parameters have no effect on bloom formation, but these effects might be very subtle and not easily controlled or distinguished from one another. Otherwise, the difference between the required fat content for sand and regular cookies is just a matter of the fat migrating more easily in the sand model. Whatever the reason, the results of our sand cookie experiments confirm that fat migration is a primary factor in chocolate chip bloom inhibition. Other parameters may ultimately be dictating how much fat is migrating during baking. Therefore, in our future experiments, we will measure the fat content and type of fat in the chocolate chip after baking and compare that with bloom formation, instead of trying to relate the initial fat content of the dough to the amount of bloom that has formed on the chocolate chips.

This leads us to our next question: If fat migration from the cookie dough to the chocolate chip during baking is primarily responsible for inhibiting bloom, how does this mechanism work?

When two fats are mixed, the resulting combination usually crystallizes much differently than either of the individual fats. This change in crystallization behavior is most likely what affects bloom formation. Milk fat (from butter) and cocoa butter are often considered compatible fats and, as such, milk fat is widely known to inhibit bloom in chocolates through its effect on cocoa butter crystallization. Although milk chocolate has less snap than dark chocolate, it is also less prone to bloom than dark chocolate. So it is not a surprise that chocolate chips in cookies made with butter are resistant to bloom. The milk fat that migrates from the cookie dough into the chocolate chip during baking protects the chocolate chips from bloom.

The vegetable fats used in shortenings are typically not compatible with cocoa butter, though, and tend to promote bloom in certain circumstances (like the surface of an old peanut butter cup that has bloomed because peanut oil has migrated into the chocolate). Despite this, chocolate chips in cookies made with vegetable shortenings also do not bloom when the fat content is high enough. Even the chocolate chips in cookies made with olive oil resist bloom (they also do not get eaten—yuck!).

Why do vegetable fats and oils protect against bloom? It is hard to say. Perhaps in the context of chocolate chips in cookies, these vegetable fats impart bloom inhibition not observed in other contexts. There could be still other factors besides fat migration that affect bloom in chocolate chips. Water content, protein from eggs, and even the natural emulsifiers found in egg yolks may affect how the cocoa butter resolidifies in chocolate chips after baking. Anything that alters cocoa butter crystallization and polymorphic transformations upon resolidification of the chocolate can influence bloom. Further studies to vary these parameters are likely to shed light on the mechanisms of bloom protection in chocolate chip cookies.

Where does this lead us? With a better knowledge of why chocolate chips do not bloom when baked in cookie dough, perhaps new methods of bloom prevention for the chocolate industry might be developed. While such a technology will not prevent your chocolate bar from melting in the car in the summer, it could prevent the chocolate from being ruined as a result.

Acknowledgment

Funding from the PMCA, an international association of candy manufacturers, is gratefully acknowledged.

Further Reading

Hartel, R., and Y. Kinta. 2010. "Bloom Formation on Poorly-Tempered Chocolate and Effects of Seed Addition." *Journal of the American Oil Chemists' Society* 87:19–27.

Lonchampt, P., and R. W. Hartel. 2004. "Comparative Review of Fat Bloom in Chocolate and Compound Coatings." *European Journal of Lipid Science Technology* 106:241–274.

Timms, R. 2002. "Oil and Fat Interactions." *Manufacturing Confectioner* 82, no. 6:50–54.

eleven

BACON

The Slice of Life

TIMOTHY KNIGHT

bacon is magical. It can transform an ordinary meal into an extraordinary delight. With just one bite, you get an irresistible crunch, a distinctive smoky flavor, and an unmistakable sense of deliciousness. This chapter takes you through the finely honed mandatory steps that turn a humble piece of pork into the mouth-watering slice of "meat candy" that we know and love. So hang on tight. You are about to embark on a journey behind the magical bacon curtain, where you will learn how a lowly pork belly becomes the meat that makes your life complete.

A Brief History of Bacon

For more than three thousand years, bacon was made on farms using traditional practices that involved salt curing, dry curing, and smoking pork bellies. During the 1770s, John Harris, an industrious farmer in Wiltshire, England, established himself as the first large-scale bacon manufacturer in the modern world by using so-called wet-curing methods. With the onset of the industrial era came Philip Armour's refrigerated rail cars and the development of more advanced preservation techniques by Gustavus Swift, both of which paved the way for the development of bacon as we know it today. In

1924, Oscar Mayer took his rightful place on the smoky-salty bacon throne by introducing the first presliced bacon. However, shoppers still had to get their bacon from the in-store butcher. So, in 1948, Mayer introduced the first prepackaged bacon, a durable cellophane-wrapped slab of sliced bacon on a thin sheet of cardboard. This allowed shoppers to select packages themselves from the retail case. In 1962, with the onset of new polymer-film technologies, Oscar Mayer began vacuum packaging his bacon (and other processed meats), once again revolutionizing retail meat packaging. An airtight envelope protected the bacon against spoilage. The back-of-package window, which allowed shoppers to see exactly what they were purchasing, was embraced in 1973.

BACON BITS
Until the sixteenth century, *bacon* was the Middle English term used to refer to almost all cuts of pork.

It All Starts with a Pork Belly

A pork belly does not come off a hog resembling anything like bacon. In fact, it is not actually the belly or the stomach; rather, it is the lean and fat from the side of the hog that remains after the ham, shoulder, ribs, and loin are removed. Each hog has two "bellies." The anatomy of a pork belly is complex, having several distinct and interspersed layers of lean and fat, which can be readily identified when looking at an individual strip of bacon. These layers are not consistently proportioned or spaced throughout the length of the belly, which is why individual slices of bacon look different (figure 18). Trimming removes sections that are too fatty to make into bacon. White bacon is made from the abdominal region of the hog and is nearly devoid of lean meat. This fatty cut is used primarily for flavoring. In some cultures, the fat from this cut is slowly melted out of the meat structure. The remaining bacon then solidifies as a crispy-crunchy "chip."

BACON BITS
Living "high on the hog" came about as a description of the wealthy because they could afford the more expensive cuts of pork that came from the anatomically high portion of the pig.

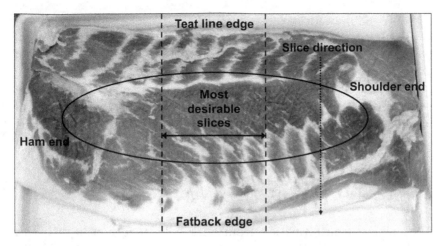

Figure 18 The lean (or internal) side of a pork belly. Note how different the slices will be as they are cut across the belly from the teat line edge to the fatback edge. The muscle is buried within the middle of the belly, and the area where it is found is represented by the oval. It is widest near the middle of the length of the belly. This is where the most desirable slices come from.

You've Got Bacon Fever—We've Got the Cure

Prior to modern refrigeration, the dry curing of pork was necessary to prevent spoilage. It was accomplished by applying a salt crust to the exterior of the pork belly, where it was absorbed over time, essentially drawing moisture out of the meat and preserving it for later use. Air drying may also have contributed to preservation. Old World dry-cure mixtures typically consisted of unpurified salts and sometimes sugar. This method worked well for preservation but the quality left much to be desired; more often than not, the result was extremely salty and harsh with inconsistent salt dispersion.

Modern bacon manufacturers use a more efficient and reliable method called wet curing, or Wiltshire curing. This evolutionary method was first routinely used by John Harris and involved submerging raw pork belly in concentrated salt brine. Injection, a modern modification of this process, uses large needles to infuse brine directly into the interior of the belly instead of relying on migration of the salt from the surface over time. It provides even curing and more effective brine incorporation. Today, most bacon brines are a mixture of salt, sugar (to lessen the harshness of the salt), and a small amount of sodium nitrite for preservation purposes.

In my opinion, quality wet-cured bacon is consistently superior to the dry-cured variety because it relies on refrigeration and vacuum packaging for preservation rather than the severe conditions that preserve dry-cured products. It is also less prone to spoilage and the development of rancid flavors and aromas. This being said, there are very good dry-cured bacons of exceptionally high quality in the retail market today. All bacon lovers should experiment with and enjoy these—and not only for special-occasion meals.

The Magician's Magic Box

The magical transformation from pork belly to bacon begins with the brine needle. After injection, the pork belly is hung to help facilitate moisture loss and prevent curling during the smoking process. As a caterpillar disappears into a chrysalis and eventually materializes as a beautiful butterfly, so too must pork belly go into its smokehouse cocoon to later emerge as delicious bacon. And as with all great magic tricks, there is a puff of smoke. In the smokehouse, pork bellies are cured, smoked, and cooked before they appear as unsliced bacon. This process can take as little as a few hours or up to a few days, depending on the manufacturer. The finished slabs are chilled, sliced, and packaged. Vacuum packing keeps destructive oxygen molecules away from the bacon, prevents the fat from becoming rancid, and helps prevent microbial spoilage. Bacon is almost always purchased refrigerated, but unopened packages can be stored frozen for months.

Up in Smoke

Smoking is a vital ingredient in the sweet science of bacon, and careful control of the delicate art of smoking is crucial to producing a quality product. Meats were originally smoked primarily for preservation, but today we are able to take advantage of smoke for much more. The smokehouse process not only partially cooks the pork but also generates the smoky flavor, aroma, and color we love and associate with bacon. In addition, this all-important step brings out the flavors in the cure and partially renders the fat. There are several methods that can be used to apply smoke to meats. The traditional and most desirable method is burning hardwood in a smokehouse. A modern method is the use of liquid smoke extracts, but it is extraordinarily difficult to reproduce the complexity and uniqueness of true hardwood-smoked meat.

When wood is heated to high temperatures, controlled oxidative combustion generates the complex mix of vapors and particulates known as smoke. Wood is made up of compounds classified into cellulose, hemicellulose, and lignin. During controlled burning, cellulose and hemicellulose break down into smaller compounds, including carbonyls and organic acids. Lignins devolve into various phenolic substances. All of these are vaporized by heat and become components of smoke. If done correctly, the combustion should result in smoke with a pleasant balance among organic acids, phenolic substances, and carbonyl compounds deposited on the pork-belly surface. The organic acids deliver a very slight tartness, but phenols provide the predominant flavor and aroma that we associate with smoke. They are fat soluble and concentrate on the exposed belly fat. Deposition of carbonyl compounds imparts a golden brown color on the lean surfaces of the bacon slab. This deposition is mostly caused by nonenzymatic browning—that is, by a Maillard reaction, in which carbonyl compounds are key (for a more incisive look at the role of nonenzymatic browning, see chapter 13).

Any bacon worth its salt should be hardwood smoked. The classification of hardwood versus softwood has nothing to do with the trees' physical hardness. It is a matter of how the trees produce their seeds. Hardwoods produce flowers and softwoods produce cones. The flavor differences between the two woods are associated with differences in the amount and types of lignins in the wood. Hardwoods are used for smoking meat because they impart desirable flavors and aromas when burned, and the smoke is sweet and clean. They include apple (slightly sweet, fruity, mild), beech, birch, cherry (sweet, fruity, mild), hickory (strong, hearty), maple (sweet), and oak (strong, earthy, robust). Softwoods produce poor-quality smoke, imparting an unpleasant flavor and smell that is described as oily, sooty, and resinous, reminiscent of the odor of cleaning solutions.

Smoke is not the only contributor to the complex flavor of bacon, but it is one of the most predominant and important components. Poor smoking will ruin your bacon. And remember, as with most things in life, it is the *quality* of the smoke, not the *quantity* that matters. True bacon artisans work with smoke like artists with paintbrushes, having perfected their craft over time. Because it is an art as much as a science, it must be perfected with experience.

Bringing Home the Bacon

Now the transformation is complete! But there is still one final part to the magic trick before the big reveal—that perfect slab in the grocer's freezer. Luckily, a savvy consumer who knows what to look for has a great advantage in selecting the best bacon. In Europe, shoppers are lucky enough to have a significant amount of bacon sold at traditional butchers, where it can be cut from a large side upon request. In North America, however, we all know the maddening sensation of rifling through various bacon packages, trying to decide which one to purchase. So, how do you choose the best bacon? While the answer depends mainly on usage and personal preference, you should look for a few key indicators.

BACON BITS

The phrase "bring home the bacon" comes from an Old English rule that says any man who could vow before God that he had not fought with his wife, or wished he were single again, for an entire year would be awarded a side of bacon.

Examine the strips for a good fat-to-lean ratio. The lean should be shades of reddish-pink and the fat should be white without any blemishes or discoloration. The bacon should have long streaks of evenly distributed lean that run nearly the full length of the slice. These lean streaks should be separated by snowy white fat. "Why does this matter?" you ask. Well, my scholarly shopper, research has shown that the most desirable bacon, in terms of uncooked and cooked appearance and palatability, is approximately 40 percent lean and 60 percent fat, with an even distribution of lean and fat throughout each strip. Slices with these proportions come predominantly from the middle section of the pork belly. This region is the most uniform part, possessing a large and lean musculature. Also, these slices will remain straighter and shrivel less during cooking. Consumer-preference research shows that the ideal width of a raw slice is approximately 1 inch (2.5 cm). At more than 1¼ inches (3 cm), the slice's fat-to-lean ratio is often too high. At less than ¾ inch (2 cm), the texture can suffer.

Next, look carefully at the color of the lean part of the bacon. During the curing process, myoglobin, the major pigment molecule in the lean portion of the meat, undergoes a series of chemical reactions that generates different-colored molecules, eventually binding with a modified form of nitrite to create the dark red pigment. This is where the pigment conversion should stop for bacon: slightly dark reddish-pink. You should see at least some lean that is this color. If the lean is entirely light pink and the red has turned to pink, it has been overly heat processed and will likely taste more like ham than bacon. You might say, "Back off, I love ham!" Well, stop your squealing! No one is saying anything bad about ham, but this is bacon, so it needs to taste like bacon, with its delicate nuances and true bacon flavor.

Once cooked, the flavor should be porky and salty, with predominantly roasted and brown top notes and a touch of sweetness, along with a clean, nonashy hardwood-smoke character. The smoke flavor should not be so powerful that it overwhelms the unique deliciousness that is generated during the curing and cooking processes. In other words, it should taste like bacon and not as though you have been slapped in the face with a slice of triple-smoked ham. If all you taste is smoke, you have been cheated out of your bacon experience. Bacon should never have undesirable off-flavors, like sourness or mustiness. It should develop a delicate crispness when cooked, and it should not have a tough or rubbery texture or seem overly greasy in the mouth.

BACON BITS

During cooking, an individual slice of wet-cured bacon normally shrinks 40 percent in length and becomes 20 percent thinner because of the loss of fat and moisture.

Crispy Matters

You have brought home the bacon. You are anticipating the perfect moment when everything becomes quiet and all you hear is that unmistakable crackling (in the pan) and crunch (in your mouth) of a perfectly cooked strip. While all basketball players know that making the perfect shot is "all in the wrist," serious bacon connoisseurs know that making the best bacon is "all in the crisp." We all have experienced how perfectly cooked bacon can take an ordinary bacon, lettuce, and tomato sandwich and transform it into a crispy oasis of delicious goodness. Now there is scientific proof of how critical

BACON BITS

Many fast-food and sit-down restaurants cook bacon ahead of orders and hold it until needed. During this holding time, the crispness is lost. Just think of the quality improvement that a restaurateur could achieve by cooking bacon on demand.

that crunch is to your lunch. The University of Leeds in England has developed an equation that demonstrates how necessary bacon crispness is to the BLT. In fact, it was determined to be the key element in the ideal sandwich. Moreover, it is the sound of the crunch that is absolutely critical to a truly desirable BLT. The researchers determined that in the best sandwich, bacon is crisply grilled and of moderate fattiness. They achieved this by cooking the bacon for 7 minutes at 480°F (250°C) over a preheated oven grill.

You Brought Home the Bacon—Now Fry It Up in a Pan

The thermal treatment applied in the smokehouse is relatively mild compared with temperatures achieved on your stovetop. Bacon needs to be cooked to a typical frying temperature to allow each and every precious slice to live up to its maximum delicious potential. When a strip of bacon is prepared, there is a loss of fat and water. This serves to concentrate flavors, promote nonenzymatic browning reactions, and generate subtle flavor, aroma, and color compounds. As more and more moisture and fat are lost from the slice, it becomes crisp and crunchy, presumably because it moves into a very stable, or glassy, state (for details about the glassy state, see chapter 24). However, if you overcook bacon, all the moisture will seep out, making the structure overly dry—and after it cools, an extremely rigid microstructure of empty cells will result in a brittle and crumbly texture. Proper browning, along with curing and smoking, combine into an explosion of deliciousness when you bite into a strip. Although pan frying is the most common way to prepare bacon, it is certainly not the only method, and many people agree that oven baking is the best way to prepare it. The oven provides greater evenness of heating and produces slices with fewer convolutions and less shrinkage. Baking is easy and results in crisp, flat, and evenly cooked strips of bacon. Just line a baking sheet with a piece of aluminum foil and place a wire cooling rack on top of it. Arrange the slices perpendicular to the tines without overlapping, and bake in a preheated oven at 430°F (220°C) until the bacon reaches the desired crispness.

Trying a variety of slice thicknesses can dramatically enhance your bacon experience. If you normally use the standard thickness, let your hair down, act with reckless abandon, and pick up a pack of thin or thick strips! Thin sliced (0.8 mm) will give you a crispy strip that easily becomes brittle. Regular thickness (1.6 mm) is very popular because it is thick enough to develop crispness but not thin enough to easily overcook and turn brittle. Thick sliced (3.2 mm) is very hearty but also more difficult to crisp. It needs to be cooked over a lower heat for a longer time to achieve a crisp slice that is not tough and chewy.

BYOB: Bring Your Own Bacon

Since it is complementary to so many food pairings, bacon does not have to stand alone on the breakfast plate. The salty flavor mixes well with both the sweet and savory, so let your imagination and taste buds run wild. It can be used as a main ingredient in dishes, such as the classic bacon club sandwich, quiche, or a breakfast casserole. Bacon, as well as rendered bacon fat, can heighten the flavor of any number of foods. You can add it to soups, stews, or roasts, and the rich smoky flavors especially complement bean dishes.

Rediscover bacon's material properties. It is flexible when raw yet holds a defined shape when cooked. Try wrapping a thin slice of raw bacon around a metal rod before baking. When crisped, it can be carefully slid off to produce a delicate spiral shape. Turn a muffin pan upside down and cover the outside of the cups with bacon strips. Broil until crisp to form a bowl shape that can be filled with any number of sweet or savory treats. You can even use raw bacon to form a lattice pattern on top of a main dish and crisp it under a broiler.

Bacon complements sweet flavors very well. Bacon cured with maple or brown sugar pairs nicely with vanilla ice cream. Strips of cooked bacon can be easily candied by quickly dipping them into exotically flavored syrup sugar, at the soft-crack stage, to form a thin glassy film. For the gourmet, a delicious and easy way to make an exciting bacon treat is to paint the bacon with maple syrup during cooking, or you can liven up the bacon by sprinkling it with brown sugar and sweet spices just before oven baking. It will also balance the sweetness of an apple pie. Just crumble cooked bacon pieces and cheddar cheese shreds on top to bring out an entirely different array of flavors. I have even seen crisp strips used as swizzle sticks for tomato-based cocktails or as desserts after being covered in a thin, delicate layer of rich dark chocolate. The possibilities are practically endless with a bit of creativity thrown in for good measure.

The Slice of Life

We have seen that bacon is an extremely versatile food; however, I will argue that we do not incorporate it into cooking nearly enough. Whether used as a breakfast meat, a cooking ingredient, or strictly for flavoring, bacon is one of the most universally enjoyed foods in the world. It is hard to deny that bacon is an important part of our most fundamental culinary experiences. It is steeped in tradition and heritage, with nearly all cultures having some form of bacon in their regional cuisines. Today, we have unprecedented access to countless varieties of bacon from around the world. There is always a new style to help satisfy our most powerful bacon cravings, from the standard English rasher to the Chinese *lop yuk*. And to think, all of this comes from a humble cut of pork that was once considered low on the hog. Embrace the pig, and give yourself over completely to the meat that makes your life complete.

Further Reading

Blumenthal, Heston. 2008. "Nitro-Scrambled Egg and Bacon Ice Cream." In *The Big Fat Duck Cookbook*, 258–263. London: Bloomsbury.

Person, R. C., D. R. McKenna, D. B. Griffin, F. K. McKeith, J. A. Scanga, K. E. Belk, G. C. Smith, and J. W. Savell. 2005. "Benchmarking Value in the Pork Supply Chain: Processing Characteristics and Consumer Evaluations of Pork Bellies of Different Thicknesses When Manufactured into Bacon." *Meat Science* 70:121–131.

Young, B., and L. White. 2008. "Bacon 101." *The National Provisioner.* Tech Journal Series. Available at www.provisioneronline.com/ext/resources/march2011images/Tech_Journals/PO-Bacon-Tech-Journal-08.pdf.

twelve

SCANDINAVIAN "SUSHI"

The Raw Story

PIA SNITKJÆR AND LOUISE M. MORTENSEN

although sushi is a new concept in Scandinavia, we have a long tradition of eating raw (non-heat-treated) cured fish. Sushi is the traditional way of serving raw fish in Japan, which spread to many parts of the world during the late twentieth century. When sushi arrived in Denmark in the 1990s, there was some skepticism because of food-safety issues but, apparently, also because of a cultural aversion to raw meat. However, most Danes forget that Denmark and the other Scandinavian countries consume great quantities of raw cured fish as well as other raw meat products. Since most Danes nowadays buy such products ready to eat in the supermarket, the consumer generally lacks knowledge of their origin and preparation. Salmon and herring are among the most popular fish eaten without any prior cooking.

In Scandinavia, herring is commonly eaten smoked and salted. Raw-cured herring is essential for the traditional Danish lunch enjoyed on Christmas, Easter, and other special occasions. Once salted, the herring is prepared by steeping it in water and marinating it in sugar, spices, and vinegar. The marinade further preserves it. Large quantities of herring are sold marinated in jars with various spices, and it is sometimes steeped in a mild curry sauce. It is served cold—straight from the jar—on rye bread with raw onions. This traditional way of serving herring is today known as "granddad sushi" among

GRAVLAX

2½ ounces (70 g) coarse salt
1¾ ounces (50 g) sugar
Freshly ground pepper
2 bunches fresh dill (6½ ounces [180 g])
3½ pounds (1.5 kg) fresh salmon with skin intact, one
   prefrozen piece*

Mix the salt, sugar, and pepper. Rinse the dill carefully and shake dry. Remove all bones from the salmon, if not already done. Cut the salmon in half and place the two pieces, skin side down, on a work surface. Rub the salt–sugar–pepper mixture into the exposed flesh of the fillets. Distribute the dill sprigs evenly across one of the fish halves. Top this fillet with the second piece of fish, skin side up. Wrap the "salmon sandwich" carefully in several layers of plastic wrap.

Place the salmon in a baking dish or something similar. Apply to the fish a weight of about 6½ pounds (3 kg), if possible (for example, books or canned goods). Refrigerate for 2 to 3 days, turning the salmon a couple of times a day.

When ready, discard any liquid, and scrape off the salt, sugar, and dill. Slice thinly with a sharp knife, detaching slices from the skin. Serve with mustard sauce, fresh dill, lemon, and bread.

The salmon keeps for approximately 3 days in the refrigerator, but it can also be frozen again.

* Eating raw or undercooked fish can put you at risk for foodborne illness. To avoid a parasitic infection, freeze the salmon before use. The U.S. Food and Drug Administration (2011) recommends freezing and storing fish and fishery products at an ambient temperature of −4°F (−20°C) or below for 7 days (total time), freezing at an ambient temperature of −31°F (−35°C) or below until solid and storing at an ambient temperature of −31°F (−35°C) or below for 15 hours, or freezing at an ambient temperature of −31°F (−35°C) or below until solid and storing at an ambient temperature of −4°F (−20°C) or below for 24 hours.

the younger generation. To make the dish complete, schnapps is served with the fish, which makes granddad even happier!

Salted and smoked salmon is yet another delicacy highly appreciated today. A pressed form of salted and fermented salmon was developed in Scandinavia in the Middle Ages, today known as gravlax, gravadlax, or lox in English-speaking countries. The word *lox* is often used in the United States when salmon is served on bagels. The terms have their origin in the Scandinavian languages. *Lax* (Swedish) or *laks* (Danish and Norwegian) is the word for "salmon," while *gravad* means "buried." In the Middle Ages, people buried the salmon in the ground, leaving time for osmotic processes as well as fermentation. The salmon was preserved with sugar and salt, both producing an osmotic pressure. Salt-tolerant lactic acid–producing bacteria further preserved the salmon by fermenting the sugar, which reduced the pH value (more about these preservation techniques follows). As time went on, the lox evolved into a lightly salted and unfermented product. Today, it is produced by sprinkling salt, sugar, and dill on the fish fillets and refrigerating them for a couple of days. To enforce the contact between the fish and the salt, sugar, and dill mixture, as well as to press out excess fluid, the fish is weighed down during cooling. The dissolved salt has a tenderizing effect on the salmon by breaking down some of the proteins. In Denmark, the raw preserved salmon is served thinly sliced on bread with a mustard sauce (figure 19). Try it yourself by following the recipe for gravlax.

The curing techniques of drying, salting, and smoking date back many hundreds of years and were invented to extend the shelf life of fish and other food. Such handling of the fish not only prolongs the shelf life but also changes the texture and flavor of the product. In the Middle Ages, raw cured fish was a staple part of the Scandinavian diet. The invention of various fish-curing techniques made it possible to store, transport, and trade fish. Cured fish became very important not only for feeding the local people but also for the great trading opportunities it provided. Today, we still use the same basic techniques, only they are slightly modified and no longer essential for preservation. What was invented due to a need to preserve the fish is now used mostly as a way to change its sensory characteristics.

The term *curing* in preservation refers to the use of antimicrobial techniques that prolong shelf life. The idea behind preservation is to make the conditions for microbial activity less ideal but also to

---

MUSTARD SAUCE

1½ ounces (42 g) dark brown sugar
½ ounce (15 mL) wine vinegar
4 ounces (120 mL) Dijon mustard
4 ounces (120 mL) sweet mustard
15 g salt
10 g pepper
6 ounces (200 mL) good-quality olive oil
Fresh dill, chopped

Mix the sugar, vinegar, mustards, and salt and pepper well. Slowly add the oil while whisking with a fork or using a handheld electric mixer.* Add the dill to taste.

* The mustard sauce is an oil-in-water emulsion just like mayonnaise. The success of this sauce relies on establishing this emulsion. An emulsion is a mixture of two immiscible liquids (oil and water) in which one is dispersed in the other. In this case, the olive oil must be dispersed (distributed in small droplets) in the water phase, which is made up of the vinegar and the water from the mustard. The plant particles from the mustard act as stabilizers in the emulsion. It is recommended that the olive oil be slowly added to the water phase while whisking. This serves to emulsify the oil droplets one by one in the water and prevent the oil from forming its own coherent phase. It is necessary to apply mechanical energy to overcome the surface tension and separate the oil into small droplets that can be sufficiently emulsified in the water phase. If more than the suggested amount of oil is added, at some point the emulsion will become unstable and separate into two phases because of the uneven relationship between the oil and water. There must be sufficient water to emulsify the increasing amount of oil. Failure to establish an emulsion when preparing the sauce results in a thin, nonuniform, separated product. Failure can be caused by incorrect proportions of mustard and vinegar: insufficient water or too much oil shifts the balance of the two phases, or oil added too quickly and not whisked enough for emulsification in the water phase can destroy the sauce. If you are already a mayonnaise expert, this will be child's play.

Figure 19 Two varieties of smoked fish: (*left*) marinated herring with curry sauce on rye bread ("grand-dad sushi") and (*right*) gravlax (cured salmon) with mustard sauce on rye bread.

slow down enzymatic reactions. Microorganisms need water to live; therefore, many preservation techniques remove water or decrease its availability. Drying and salting are examples of very old techniques that deprive food-borne microorganisms of water. Smoking does the same to some degree, provided it is carried out with heat, which encourages some water to evaporate. In addition, smoking makes life hard for the microorganisms because of the high temperatures and certain chemical components of the smoke that are taken up by the food. Another way of making life difficult for the microorganisms is to control the pH value. Since most microorganisms do not like an acidic environment (low pH), marination in acetic acid (the principal acid in vinegar) or fermentation by lactic acid–producing bacteria are other common preservation techniques.

Several of these techniques are often combined in order to give an additional preservation effect as well as the desired flavor and texture. It is important when preserving food to minimize the oxidation of the fat, which lends a unpalatable rancid flavor to fatty products. The preservation may, however, accelerate oxidation since oxidation is induced by sunlight and the rate of the chemical reaction is increased at very low water activities as in dehydrated products. The high levels of unsaturated fatty acids in fish (like salmon or sardines) render it highly vulnerable to the development of rancid off-flavors.

In ancient times, people removed water simply by exposing the food to sun and wind. To prevent bacterial growth, the water content must be reduced to 25 percent and further to 15 percent to prevent fungal growth. The texture of the fish changes during drying because the structure of the meat is altered. This alteration causes the formation of new flavors initiated by enzymatic reactions. Moreover, dehydration results in a simple concentration of certain flavor molecules, which contributes to the changes in flavor during drying.

In the dark and humid Scandinavian climate, sunlight is not a reliable source for drying food, so fish easily spoils before it dries sufficiently. Salting is another ancient preservation technique—also combined with drying or smoking—that preserved fish independent of the weather conditions. Preservation of fish by salting can be traced back to ancient times, from between 4000 to 3500 B.C.E., having reached its peak in the eighteenth and nineteenth centuries. The availability of salt was essential, and large establishments were necessary to secure the required salt supplies.

When salting food, water migrates outward because of a high concentration of dissolved solutes on the food's surface. This phenomenon is known as osmosis and results in a drying effect. So-called osmotic pressure can be applied using other solutes, like sugar, although a much higher sugar concentration is needed to obtain a similar effect. The efficiency of salts in osmotic dehydration comes from their electric charge. During the salting process, salt will diffuse into food products, and the dissolved salt binds the water, thereby further lowering the availability of water and with it the water activity. The salt will also cause the microorganisms themselves to become dehydrated, making them unable to grow and reproduce.

Various salting techniques have been invented. One example is immersion in brine. It is an especially advantageous technique for fatty fish because it protects against oxygen-induced rancidity and also promotes contact between the salt and the surface of the fish. Another advantage to using brine is that the salt is already dissolved, which enhances its penetration. Applying pressure during salting is another way of promoting the penetration of salt (see the recipe for gravlax). A technique used for the processing of herring, invented around the thirteenth century by the Dutch and northern Germans and still used today, involves the removal of the gills, gut, and stomach, while leaving some portion of the enzyme-rich intestines. The enzymes induce the so-called ripening process, which gives a tender texture and adds flavor to the product.

The ripening process involves some complex physical and chemical changes in the fish. Enzymes are responsible for the transformation of proteins, lipids, and carbohydrates, resulting in the particular flavor and consistency of salted fish. Although a salt concentration exceeding 5 percent kills most microorganisms, some salt-tolerant bacteria remain and play a critical role in producing the changes in texture and flavor characteristic of salting.

Salting has long been an efficient way of extending shelf life, and changes in the taste quality of foods (for good or bad) were a necessary consequence. Because we understand the process better today and because we are not dependent on the preservation effect, we can use salting to achieve desired changes in the flavor and texture of a product.

It is probably the case that smoking was invented because it was necessary to dry food close to the fire. In the low-lit and humid Nordic countries, the heat from fire was the only way of drying fish and other foods, especially in the winter. As mentioned, smoke has some advantages over drying alone because some of the smoke particles, such as phenolic compounds, have an antimicrobial effect when deposited on food. Moreover, some phenolic compounds have an antioxidative effect, that is, they reduce the formation of rancid off-flavors from the oxidation of the unsaturated fats in fatty fish. In addition, the food generally loses some fat during the smoking process, which is then no longer available for oxidation. The smoke adds many desired flavor compounds (phenols and furan derivatives) to the food that may mask other off-flavors, thereby increasing the gastronomic value of the food. Reactions between sugar and amino acids impart color to the fish during smoking. These reactions are collectively known as the Maillard reaction, which is responsible for flavor formation (for an excellent account of the Maillard reaction, see chapter 13). Heat denaturation (hot smoke) and protein degrading, so-called proteolytic enzymes (cold smoke), lend smoke its tenderizing properties. With the introduction of more effective preservation methods, the early function of smoking lost its importance and instead the unique sensory properties of smoked foods became the focus.

Smoke is a highly appreciated flavor in Denmark as it is in other countries. Even so, only the most enthusiastic foodies have taken up the smoking of food at home. Everyone else can get smoked products in the form of bacon, fish (mackerel, herring, salmon), and fresh cheese in any Danish supermarket. Smoked herring has become a tourist attraction in itself on Bornholm, one of the many Danish islands. Although smoking has never gone out of fashion, it has been experiencing a resurgence in recent years. This is because of the growing interest in Nordic traditions and ingredients among Scandinavian chefs who want to create a new Nordic cuisine. Hence, many popular Danish chefs have taken up this technique, and new smoked delicacies are appearing on menus in high-end restaurants across Denmark.

Although the tradition of eating cured fish is still strong in Denmark and the other Scandinavian countries, Japanese sushi has become accepted and is generally considered a healthy and delicious food. Danish supermarkets sell tools for making sushi, and school canteens offer it for lunch. In Denmark, sushi is often served at celebrations for its colorful and attractive look. Moreover, a new term, *smushi*, which is a combination of *smørrebrød* (Danish open sandwiches made with rye bread) and *sushi*, now appears on the menu in several cafés and restaurants. A smushi serving is a selection of small tapas-like open sandwiches, such as herring on rye bread (granddad sushi) and gravlax on bread.

Although the Danish food culture and cuisine are very dynamic, always changing under the influence of other cuisines, Danes do not easily let go of traditional foods. They love them because they have acquired a taste for them from childhood and because they connect them with celebrations. The Danes' early need for preserving food has been of great importance to several well-known and appreciated fish preparations in Denmark today. It is interesting to notice how the Japanese fish preparation, sushi, is both similar to and different from the Scandinavian serving of cured salmon and herring. Both use local fish. Both are served raw (non-heat-treated), sometimes fermented, with the local cereal staple (rice in Japan, rye in Scandinavia) and with a pungent accompaniment (wasabi with sushi, mustard with gravlax, and raw onion with herring). Despite the similarity, the outcome is markedly different.

Personally, we appreciate traditional dishes, served over and over again, just as much as we appreciate the constant arrival of new foods and flavors from various parts of the world. We sometimes wonder what our grandchildren will serve us for dinner. Will our traditional gravlax and marinated herring survive at Christmas lunch or will they be superseded by new and exciting foods? We can only hope that the continuous evolution of our cuisine never stops so there will be plenty of surprises on the table when we grow old!

Further Reading

U.S. Food and Drug Administration. 2011. "Parasites." In *Fish and Fishery Products Hazards and Controls Guidance*. 4th ed. Available at http://www.fda .gov/Food/GuidanceComplianceRegulatoryInformation/Guidance Documents/Seafood/FishandFisheriesProductsHazardsandControlsGuide /default.htm.

thirteen

MAXIMIZING FOOD FLAVOR BY
SPEEDING UP THE MAILLARD REACTION

MARTIN LERSCH

an idea that struck me once was to add baking soda
to browning onions. I chopped an onion, melted butter in a frying
pan, and added the onions together with a pinch of baking soda. And
voilà (as Louis-Camille Maillard himself would have said): the color
of the onions changed faster than without the baking soda. The taste
of the browned onions was remarkably sweet and caramel-like, and
compared with conventionally browned onions, they were softer—
almost a little mushy. By the addition of baking soda, I had changed
the outcome of an otherwise trivial and everyday chemical reaction,
and the result seemed interesting from a gastronomic perspective!

The idea of the baking soda addition was not taken out of the blue
but based on something I gleaned from the chemistry of the Maillard
reaction. Popularly known as the "browning reaction," the Maillard
reaction is the chemical interplay between a reducing sugar (a sugar
that, under alkaline conditions, forms reactive ketones or aldehydes)
and an amino acid (the basic building block of all proteins). As a
chemist, I have always found the Maillard reaction to have a decep-
tive name, camouflaging the fact that a surprisingly large number
of reactions occur when a reducing sugar and an amino acid are
heated together. In addition to its complexity, I had noted the pH
dependency of the Maillard reaction. By increasing the pH—making
the food less acidic and more alkaline—the Maillard reaction can be

sped up. And the addition of baking soda happens to be a convenient way of doing this. Over time, it became clear to me that the use of baking soda was only one of many ways cooks can and do influence the speed of the Maillard reaction in the kitchen.[1]

Ever since the French chemist Louis-Camille Maillard studied the metabolism of urea and kidney illnesses and published his thesis on the actions of glycerin and sugar on amino acids in 1913, the Maillard reaction has been a hot research topic. A review on browning reactions in dehydrated foods, which appeared in the first volume of *Journal of Agricultural and Food Chemistry*, remains the most cited paper in that journal's history. It is sad, yet understandable, that undesirable occurrences of the Maillard reaction have received more attention in the scientific community than the desirable ones. Fortunately, desirable Maillard products have been explored for thousands of years in the kitchen and the results are well documented in numerous recipes.

The Maillard reaction, which is also sometimes referred to as "nonenzymatic browning," produces volatile compounds that contribute aroma and nonvolatile compounds that provide color, known as melanoidins. Some of these compounds contribute to the resulting flavor as well. The Maillard reaction imbues foods with a characteristic smell, taste, and color. High-temperature processes in particular, such as frying, roasting, grilling, and baking, rely heavily on the Maillard reaction for the characteristic aromas it produces. What would the crust of a freshly baked loaf of bread be without the Maillard reaction? What would beverages such as espresso, hot chocolate, and Irish stout be if the coffee, chocolate beans, and barley were not roasted to facilitate the Maillard reaction? Or the nice meat flavors of a beef roast? Or the smell of toasted white bread? Browned onions? The list is endless.

It is possible to speed up the Maillard reaction by choosing favorable conditions. Chances are, you have done this without knowing or thinking about the chemistry. One can speed up the reaction by

---

1 A word of warning: most foods are acidic, and this provides natural protection against the growth of microorganisms in general and *Clostridium botulinum* in particular, the latter producing *botulinum* toxin (which gives rise to botulism, a paralytic and fatal illness) in foods that are stored in the absence of air and have a pH greater than 4.6. The addition of alkaline reagents, such as baking soda or soda lye, will reduce the acidity and hence diminish the natural protection against microorganisms. Because of this, care should be taken and baking soda should not be used in marinades, for instance, that will remain in contact with meat for more than a couple of hours. The use of alkaline reagents for foods that are cooked immediately, however, is unproblematic.

adding protein or a reducing sugar, increasing the temperature, using less water (or boiling off water), or increasing the pH. In fact, when looking for examples, I was surprised at the extent to which conditions favoring the Maillard reaction had found their way into recipes.

When I was little, I remember my mother brushing leavened yeast buns with milk or egg yolk to give them a nice brown crust in the oven. She knew nothing about the Maillard reaction, but she did know how to obtain the desired color and aroma. In the glazing of baked goods, milk or eggs provide the protein source that leads to Maillard reaction browning. In recipes in which eggs are used because of their binding and emulsifying properties, the role they play as a protein source for the Maillard reaction is sometimes overlooked. An added benefit of the egg yolks when applied to yeast buns is that the viscosity allows a thicker layer to be brushed onto the surface, yielding a glossy finish. Milk provides the reducing-sugar lactose in addition to protein, which compensates for the lower viscosity with regard to browning potential.

Yeast buns can also be brushed with sugar water before baking. Even though sucrose is not a reducing sugar, it easily breaks up into fructose and glucose when heated, and these take part in the Maillard reaction. When a sugar is applied to a surface that is exposed to heat, there will be a fine line between caramelization, which involves only sugars, and the formation of Maillard products. If the surface contains proteins or amino acids, both caramelization and Maillard products will be observed. This will also be the case for the yeast buns. Another example is glazed meat, such as ham, in which the sugar reacts with proteins in the meat. Barbecue marinades and sauces for basting or brushing can contain a lot of sugar. This encourages quick browning, but it can be a disadvantage if the meat is cooked at a high temperature or for a long time. With plenty of sugar present, the Maillard and caramelization reactions proceed fast but may also go too far, yielding higher concentrations of the Maillard products and an unpleasant burned flavor. When grilling with direct heat from hot coals, it is advisable to leave the sugar out of the marinade and save the sugar-rich sauces for a last-minute brush.

Because both a reducing sugar and protein are required for the Maillard reaction to occur, the preparation of butterscotch, caramel candy, and toffee each represents a nearly perfect setup. The making of plain caramel starts with water and sugar. The water stabilizes the temperature as it evaporates and cools the syrup. This allows the syrup to be cooked for a longer period of time without burning. In this process, rich caramel flavors develop. In the making of butter-scotch, caramel candy, and toffee, butter and/or milk are added to the syrup. This provides the required proteins for the Maillard reaction to occur together with the caramelization.

In several countries, including Spain, Argentina, and Singapore, it is common practice to add sugar to coffee beans in the roasting process. The resulting coffee is known as torrefacto or torrado, not to be confused with torrefied coffee, which refers to conventionally roasted coffee. Several explanations exist for why this is done, including the formation of a thin sugar film to protect the beans from oxidation as well as to compensate for weight loss from evaporation (in some countries, up to 20 percent sugar is added, and sugar is cheaper than coffee!). Others claim that it is a simple way of masking the flavor of inferior beans, especially cheap robusta beans. As the sugar is heated, it caramelizes, and the sugar solution penetrates into the coffee beans, taking part in the Maillard reaction. Despite the obvious potential for less honest coffee roasters, the torrefacto method is used with success to obtain a special aroma, and it is not uncommon to find a fraction of torrefacto beans added to conventionally roasted beans. This influences the resulting flavor, emphasizing toasty, earthy, and musty flavors.

Apart from adding proteins and reducing sugars, there are other ways to influence the Maillard reaction. Temperature is crucial, and the correlation between temperature and browning is obvious. In order to obtain sufficient Maillard products within minutes or hours, a temperature of more than 212°F (100°C) is required. This is easily achieved in processes such as frying, roasting, grilling, toasting, flambéing, and baking. A typical temperature range of 230 to 340°F (110–170°C) is often cited as ideal for the Maillard reaction to proceed in the normal time frame. If the temperature gets too high, bitter flavors develop, even before the surface appears burned. If the temperature exceeds the typical range for the Maillard reaction, it is common to talk about pyrolysis, which can be characterized as heat-induced decomposition. If uncontrolled, pyrolysis of foods will typically give rise to burned and bitter flavors. However, the desirable smoky flavor in barbecue sauces and Scotch whisky comes from the controlled pyrolysis of wood and peat, respectively.

Even though the temperature is ideal for the Maillard reaction to proceed on the surface of a steak, for instance, the great challenge is that the interior of the steak should not exceed 122 to 150°F (50–65°C), depending on consumer preference. This leaves a relatively narrow window in which the temperature gradient through the steak is at the desired core temperature and sufficient Maillard products have been formed on the surface. With the *sous vide* cooking technique, this is solved by bringing the whole piece of meat to the desired core temperature in a temperature-controlled water bath, followed (or preceded) by a quick browning of the surface, either in a sizzling hot pan, on a hot grill, over a gas flame, or with a blowtorch.

Contrary to popular belief, the Maillard reaction will also occur at lower temperatures. In vintage champagne, autolyzed (inactive) yeast and sugars react to form Maillard products that yield a characteristic flavor profile. This reaction takes place in the cool chalk cellars of the Champagne district in France, in which the temperature remains constant at 48 to 54°F (9–12°C) year round. Because of the low temperature, a much longer reaction time is needed, so the characteristic Maillard-influenced flavor is found only in aged champagnes. If the temperature is increased, the reaction will proceed more quickly. When liquids such as stock or demi-glace are boiled, plenty of Maillard products are formed within hours. Similarly, a roux is cooked not only to remove the flour taste but also to allow the development of flavors. To make dark stocks for brown sauces, the meat and bones are roasted prior to boiling in order to create an even more intense meaty flavor.

The presence of water limits the maximum attainable temperature as it boils off from the surface of foods, thereby slowing the Maillard reaction. However, once water has evaporated, for example, in a bread crust or on the surface of a french fry, the drier surface allows the temperature to exceed 212°F (100°C), which in turn drastically speeds up the Maillard reaction. Similarly, a piece of toast browns in the outermost layer only. But less water is not always better. There is an optimum water level required for the Maillard reaction to proceed. If the food gets too dry, the lack of water will actually slow down the Maillard reaction as the mobility of the reagents decreases.

Another way of influencing the Maillard reaction (and perhaps the least obvious) is by adjusting the pH. The Bavarian pretzel is an extreme example of how the Maillard reaction can be tweaked, and it seems it was a serendipitous discovery. On February 11, 1839, the German baker Anton Nepomuk Pfannenbrenner unintentionally used the lye (sodium hydroxide, or caustic soda) intended for the cleaning of his baking sheets instead of sugar water to glaze his pretzels. The customers, who were used to sweeter pretzels, liked the new taste, and to this day, Bavarian pretzels and even the ubiquitous pretzel sticks are sprayed with (or immersed in) a 1 to 3 percent solution of sodium hydroxide before baking. The high pH speeds up a bottleneck in the Maillard reaction and the result is a delicious savory snack with a shiny brown finish.

A more common basic ingredient found in most kitchens is baking soda (sodium bicarbonate). Its most familiar use is as a leavening agent, which requires the addition of an acid to function. Since it is a weak base, it can be used to increase pH and hence the speed of the Maillard reaction. When making pretzels at home, baking soda can easily be substituted for sodium hydroxide. Since baking soda is a weaker base, some recommend using boiling water when immersing the pretzels (as opposed to cold water when using soda lye). When baking soda is used as a leavening agent in cookies, a side effect is more rapid browning and a more pronounced nutty flavor.

Dulce de leche is a popular sauce and caramel candy in Latin America. It is made by slowly boiling sweetened milk. Baking soda is not a required ingredient but is often included. The baking soda gives dulce de leche a darker color and contributes to the flavor by facilitating the Maillard reaction. Similarly, it is the baking soda that gives persimmon puddings their dark brown color and rich flavor. The kinds of chemical reactions observed in champagne, stocks, and caramel candies belong to the less frequently encountered examples of Maillard reactions that occur in the interior of foods. The reason the Maillard reaction primarily occurs on the surface of foods is of course because of the higher heat and lower water content (from evaporation) encountered there.

Microwavable pies with browning crusts are challenging to produce because microwaves primarily interact with water and therefore bring the temperature only up to the boiling point. This is the reason microwave cooking in general does not contribute much flavor to dishes and why microwave ovens are used mainly to reheat food. In order to get a nice browning of a pie crust in a microwave, pH adjustment is combined with the addition of reducing sugars and amino acids. Another example of baking soda use on surfaces is in Chinese and Japanese tempura batters. In addition to a leavening effect, the baking soda also gives a more rapid browning.

At the beginning of this chapter, I mentioned how a pinch of baking soda could influence the browning of onions. The browning proceeds faster and the result is a remarkably sweet flavor with strong caramel notes. The alkaline baking soda increases (or at least stabilizes) the pH of the onions, which release acidic compounds when chopped and subjected to heat. More water is lost than without the soda, and the chopped onions collapse to a certain degree. If too much baking soda is used, the onions turn mushy and wet. One possible explanation for this is that the alkalinity facilitates onion cell-wall destruction, resulting in the rapid release of the intracellular juices.

Interestingly, some recipes recommend adding salt when sautéing onions, and salt facilitates osmosis, which draws water out of the cells. The evaporation of this water adds to the overall cooking time, which may increase the amount of Maillard products. But more important, salt will of course also act as a flavor enhancer.

To simultaneously compare the effect of salt and baking soda, I chopped a couple of onions, put them in a hot frying pan with some oil, and split the onions into four equal portions. To three of the portions, a pinch of baking soda, salt, and a baking soda–salt mixture were added, respectively. The last portion served as a control. The experiment revealed a significant difference between the baking soda and the salt. With the baking soda, a faster browning was observed, and the onions came out very sweet, with caramel notes. The salt had no significant effect on the browning but did enhance the savory flavor. Also, the onions with salt retained a slight acidity that could not be detected in the baking soda portion. The onions that were browned with the baking soda–salt mixture (1:1) had the best flavor, probably due to the enhanced savory taste from the salt combined with the rich caramel sweetness.

Apart from the effect on the overall speed of the reaction, changing the cooking conditions also favors other reaction pathways, which in turn result in different flavors. For instance, in a model study, it was found that the formation of 2-furaldehyde (almondy, woody, sweet aroma) was favored at a low pH, whereas furanone (caramel-like aroma) was favored at a higher pH. The latter fits well with the observations from the onion experiment. But because of the complexity of the Maillard reaction in real food systems, there is reason to believe that much remains to be discovered about how pH affects flavor.

So far, I have discussed how the Maillard reaction can be made to proceed faster, but sometimes the opposite is desired, especially in industrial food preparation. In dehydrated products, such as instant potatoes, milk powder, egg powder, corn starch, cereals, and fruit, the Maillard reaction causes deterioration of the food colors and decreases the nutritional value. And ever since the discovery of high levels of acrylamide in fried and baked foods in 2002, a real effort has been made to reduce these levels. In home cooking, a motivation for

Table 1 Conditions That Speed Up or Slow Down the Maillard Reaction

	Speed up Maillard reaction	Slow down Maillard reaction
Protein	More	Less
Reducing sugar	More	Less
Temperature	Higher	Lower
Water	Less	More
Cooking time	Longer	Shorter
pH	Higher	Lower

slowing down the Maillard reaction could be a desire to emphasize the intrinsic flavors of the ingredients used.

The conditions that speed up the Maillard reaction can be reversed to achieve the opposite result (table 1). Using a lower temperature and a shorter cooking time is so obvious that one would not even think of it as a way of reducing the amount of browning. When cooking jam, the cooking time is kept short in order to reduce the decomposition of pectins and the formation of unwanted Maillard products. Similarly, removal of milk solids is what allows clarified butter to be heated at higher temperatures than normal butter. When making ghee (India's clarified butter), however, these milk solids are allowed to react for some time before they are removed, giving ghee its characteristic nutty flavor. The addition of water can help to lower the temperature and halt the Maillard reaction, which is what happens when a pan is deglazed with water, stock, or wine. The water stops the reactions and helps collect the flavor molecules. Excessive browning in cookies can be avoided by the addition of an acid that lowers the pH.

To conclude, it is fascinating to consider how well the Maillard reaction—in many cases, without knowledge of the basic science behind it—has been manipulated by home cooks everywhere. By adjusting simple parameters, such as sugar, water, and protein content, temperature, and pH, the Maillard reaction can be made to proceed faster or slower and therefore can influence the reaction pathway and the relative concentrations of the resulting flavor compounds. In an educational setting, this can be used to illustrate basic chemical reactions. For home cooks, it demonstrates that they may know more chemistry than they are aware of. And for the scientist, it may serve as inspiration for further study of the Maillard reaction in gastronomy. But, most important, in the everyday kitchen, this knowledge can be used by the creative cook to improve old dishes and invent new ones.

Further Reading

Association L. C. Maillard. http://www.lc-maillard.org. [The historically inter-
    ested will find much information about Louis-Camille Maillard (including his
    original thesis) at this Web site]
International Maillard Reaction Society (IMARS). http://www.imars.org. [This
    society organizes the International Maillard Symposium and maintains an
    informative Web site]
Nursten, Harry. 2005. *The Maillard Reaction: Chemistry, Biochemistry, and
    Implications*. Cambridge: Royal Society of Chemistry.

fourteen

LIGHTEN UP!

The Role of Gases in the Culinary Experience

MATT GOLDING

speaking from experience, food scientists can make for unfortunate dinner party guests, having a tendency to draw excessive and unwanted attention to the structure, composition, and nutritional value of the foods being served. Hosts tend not to appreciate explanations, no matter how scientifically informative, as to why a particular dish or course did not turn out as expected. Yet, without the application of scientific disciplines such as material science, microbiology, and processing, there would not be a modern food industry capable of feeding millions of people. While this fact admittedly does not stop some of us from being food bores from time to time, it does impart an ingrained interest in the design and properties of foods.

An example of this was provided when my family and I were living in Australia in 2007. We had had a few friends around for a traditional barbecue. There was not actually anything wrong with the meal itself; however, in cleaning up afterward, it was clear that some parts of the meal had fared less well than others. Leftover champagne and beer had long since gone flat, and the whipped cream and ice creams that had been used in the Pavlova (a New Zealand meringue-based cake) had collapsed into thin, almost watery, messes, although the meringue itself was still holding its shape. These foods and drinks, you might notice, share a common compositional feature that is essential to their culinary character—gas. In fact, food gases have a huge impact on the textural properties and physical stability of many of our most desirable foods and drinks. A question arises as to how something as nebulous as air or gas can be contained and structured so that it lends texture to food. The answer is through the creation of foams.

Foams are a class of materials, termed colloids, in which two incompatible phases, water and air, are mixed together by dispersing one phase (air) into the other (water). Emulsions are another example of colloids in which oil is mixed with water by dispersing it in the form of small droplets, typically smaller than 1 micron (about 0.00004 inch [0.001 mm]) in diameter. Such systems are generally unstable. In other words, over time, they tend to revert back into their separate states through various mechanisms. The dispersed state in foams is in the form of bubbles. As such, foams are often particularly short-lived systems and can be rapidly destabilized. The lifetime of foam is affected by several properties, most notably composition of the bubble surface, bubble size, and the physical properties of the phase surrounding the bubbles, such as stiffness, thickness, and solidity.

Let us look at how foam lifetime can be controlled through these different aspects, and how it can in turn be used in the design and production of some remarkably diverse gas-containing foods, not least our aerated dinner party dishes. Moreover, we can prepare food foams whose lifetimes extend from seconds to years.

Bubbles from Seconds to Minutes

For most foams, air is initially incorporated through some form of agitation, such as whipping, although sparging (to agitate by compression), injection, and decompression can also be used to disperse air or other gases. You can see that without any additional stabilization, such bubbles are highly unstable and tend to rapidly coalesce. That is, two neighboring bubbles will fuse together through mutual contact, or they will burst upon reaching the surface of a container exposed to air. Conversely, when bubbles do not readily coalesce or burst, it means the bubbles are stabilized by one means or another. This provides a relatively simple means of testing the purity of water: pure water is completely incapable of providing bubble stability. Therefore, you should be more than a little concerned if you find your drinking water is able to support a foam when you shake it.

Some aerated foods are characterized by their lack of foam stability. Let us consider the glass of champagne served at our summer dinner party. During the manufacture of champagne, carbon dioxide produced by the fermentation process is solubilized under pressure in the bottle itself. On opening, the concurrent pressure drop encourages the carbon dioxide to come out of solution in the form of bubbles. These bubbles are released from the liquid on dispensing and during consumption through a process termed nucleation. The rapid formation of foam observed when champagne is poured into a glass is followed by an almost equally fast collapse of that foam as the bubbles quickly coalesce and burst. Likewise, the rapid formation and collapse of bubbles in the mouth is responsible for the sensation of effervescence when drinking the champagne. Prolonging bubble lifetime by improving the stability of the foam would, in this case, have unwanted consequences. In the mouth, there would be no immediate bubble collapse, and consequently, the sensation of effervescence would be replaced with more of a frothing mouthfeel that would make the drink more difficult to swallow. Even pouring the drink into a glass would become more difficult because excessive foaming and stability would throw off the foam-to-liquid ratio. You can observe this at home when making an ice cream soda, or float (called a spider in Australia and New Zealand). The float is a combination of a scoop or two of ice cream with a carbonated drink, most commonly cola or lemon soda. The protein in ice cream serves to increase the stability of the bubbles when they form, and so the mixture tends to be a lot frothier in appearance and texture. When making one of these, it helps to add the ice cream to the drink, rather than the other way around, to avoid too much foaming!

However, properly controlled foaming for certain other alcoholic beverages is a mark of the drink's quality. The most obvious example is a glass of beer, in which the formation of a foamy "head" on the top of the glass is a highly visual marker of the beverage's quality. Ideally, the foam should last for the time it takes to drink the beer. This means, depending on relative thirst, anywhere from a few seconds to a few minutes. Most beer foam bubbles are formed in the same way as those of other carbonated beverages, with the release of a gas (usually carbon dioxide) from solution. Pouring from a bottle or can into a glass encourages release of the carbon dioxide from the liquid due to nucleation of bubbles by the rough surface of the glass. Consequently, a dry glass will cause more initial foaming than a wet glass.

Formation of beer foam requires significant bubble stabilization given how long the foam is expected to last. This can be achieved by adherence of specific molecules to the nascent bubble surface so as to create a film between neighboring bubbles, which acts as a barrier to coalescence. Molecules that act in this way are considered surface active, or "amphiphilic." There are many surface-active molecules capable of temporarily stabilizing foams. Edible amphiphiles include proteins, such as those from milk, eggs, and gelatin, as well ingredients derived from oils and fats. The common feature of all these materials is the two types of molecular regions: those that are attracted to the surrounding water (hydrophilic) and those that are not (hydrophobic).

In foams, the molecules will partition themselves at the bubble surface such that hydrophilic regions are preferentially exposed to the water phase and hydrophobic regions are more closely associated with the air phase. In the case of beer foam, the bubbles are stabilized by a number of surface-active molecules naturally present in the beer, including proteins from the malt and their by-products, as well as bitter acids from the hops. These molecules are able to quickly adhere to bubble surfaces when carbon dioxide is released from the beer during pouring, providing initial rapid-foam stability that allows a "head" to develop, and generally this head of foam is maintained during the drinking experience.

However, the bubbles in beer can be more easily destabilized than one might think. Fats and oils, in particular, are known to be particularly effective at collapsing foams, as they interfere and compete with the mechanisms of bubble stabilization in beer. Dirty beer glasses (particularly lipstick!) and the consumption of potato chips (crisps), peanuts, and other fatty snacks are often the main culprits in the premature collapse of beer foam.

Bubbles from Hours to Days

One common characteristic of the beverage-based foam is the partitioning of the aerated foam into a separate layer at the top of the drink. This maximizes the visual appeal of the foam and contributes to its texture as it is swept along with the liquid during drinking.

For these liquid-based products, it is perhaps not surprising that this should be the case. There is a considerable density difference between water and air, a factor of one thousand. Consequently, even the relatively small number of bubbles produced upon pouring a glass of Guinness will rapidly rise due to buoyancy effects, resulting in the "creaming" of bubbles under gravity to form a separate layer at the top of the glass. In the case of many other foam-based foods, the specific textural contribution of the foam requires that the bubbles are uniformly dispersed throughout. How then to stop the foam from separating as it does for beverages? The answer is to change the thickness of the liquid phase as a means of immobilizing the bubbles. Thickening the liquid—or better, making a gel out of the liquid—will keep the bubbles apart and prevent them from rising to the top. By thickening the liquid, we can then extend foam lifetime beyond just a few minutes.

A nice example of this is the whipped cream that is used as part of the Pavlova. Usually, more than half of the volume of whipped cream is taken up by air, which imparts a light, fluffy texture to the cream that melts in the mouth. To achieve this particular texture, the air needs to be evenly distributed throughout the cream.

When preparing dairy-based whipped cream at home, whipping cream with a fat content of between 30 and 40 percent is often used (it becomes increasingly difficult to produce a stable foam at lower or higher fat contents). Air is incorporated by thorough and vigorous whisking of the cream, initially generating relatively coarse air bubbles. In the early stages of whipping, protein in the cream helps to stabilize the bubbles.

As whipping progresses, however, there is a noticeable thickening of the cream to the point where the foam becomes self-supporting, allowing it to retain its shape on standing. These changes in the structure of the cream are due to alterations in the behavior of the fat globules during the whipping process. As more air is incorporated into the cream, more fat globules come into contact with the surface of the bubbles, creating a shell of droplets around each bubble. At the same time, fat droplets partially fuse with one another. This creates a rather strong network of partially fused fat droplets that greatly increases the thickness and rigidity of the cream, to the point where the air bubbles are effectively trapped within the network of fat droplets and the foam becomes self-supporting (and can remain so for several days if stored under the right conditions). Overwhipping leads to complete disruption of the membranes around the fat droplets, leading to overaggregation and the formation of fat grains or lumps. The structure of the whipped cream cannot be recovered at this point, although butter is not far off in this process, so the effort may not be entirely wasted!

Bubbles from Weeks to Years

Getting air or gas into food is not really a problem—keeping it there is another matter entirely. Of course, for many aerated foods, foam lifetime does not need to endure for more than a few hours, and yet we have some wonderful examples of aerated foods that can retain their structure and stability for considerably longer than this, extending into years. This is exemplified by the meringue component of Pavlova. Compositionally, meringue is a remarkably simple food—a combination of egg whites and sugar, with, occasionally, lemon juice, cream of tartar, or salt added. As in whipped cream, the foam is created using whisking or beating. The slightly viscous nature of the egg whites allows for easy incorporation and retention of air during beating. Concurrently, during the aeration process, proteins from the egg whites adhere to the surface of the bubbles, where they serve as a highly effective barrier to coalescence.

Many cookbooks say that the egg whites should be whisked into so-called soft peaks. At this point, there is sufficient air incorporated into the foam for the structure to become self-supporting. But it is the cooking that assures the meringue's long-term stability. Oven temperatures of 194 to 250°F (90–120°C) are excellent for drawing off moisture from the foam. In the end, the dehydrated protein forms a rigid, glassy state (for more on this concept, see chapter 24). The solidification of the protein–sugar phase in this way permanently traps the bubbles in a rigid matrix (for details on dehydrated meringues, see chapter 15). A similar action occurs in sponge-type foods, such as breads and cakes, in which the high elasticity of the water phase serves to immobilize the air phase, preventing foam collapse. The sugar in meringues, in addition to lending sweetness, is known to contribute to the material properties of the

MICROWAVABLE MERINGUES

1 egg white
10 ounces (300 g) powdered (icing) sugar

Lightly whisk egg white and stir in the sugar until mix is stiff and pliable. The mix can be rolled into small balls and then placed into the microwave (three at a time). Set microwave to high and cook for 1½ minutes. Meringue structures will rise and set during cooking.

OVEN-BAKED MERINGUES

4 large egg whites
4 ounces (115 g) superfine (caster) sugar
4 ounces (115 g) powdered (icing) sugar

Preheat the oven to 212°F (100°C). Egg whites should be carefully separated from the yolks and whisked until they form stiff peaks. Gradually add sugar until fully mixed. Spoon or pipe onto a baking sheet and cook for about 90 to 120 minutes.

Recipes from www.bbcgoodfood.com

meringue. It is even possible to make a meringue in a microwave, although here you need about ten times the amount of sugar to egg white ratio to "set" the structure.

Finally, an old wives' tale implies that beating egg whites in a copper bowl will improve the stability of a meringue. Science has provided data that suggests an interaction between conalbumin, one of the two major proteins in egg whites, and copper ions that indeed leads to improved stability of egg-white foam. However, how this actually happens still remains to be unraveled. If you are an impatient reader, chapter 15 offers more technical detail about how meringues work.

Final Frothy Thoughts

These few examples of foams from our dinner party show the impact air and other gases have on the culinary properties of many of our favorite foods and drinks. A lot of these foamed foods have originated from traditional recipes and can be traced back centuries. Their popularity has seen them subsequently adapted by food manufacturers for general consumption. The food industry enjoys working with air because it is, essentially, a free ingredient. Many consumers likewise enjoy air: it contains no calories and, as we have shown, creates some wonderful textural experiences. However, creating and stabilizing foams with perfect consistency, whether in the kitchen or in the factory, is neither easy nor trivial. From the perspective of a food scientist, this is good news, as we are continually challenged to improve the stability and quality of foods, whether structured with air or other food components. In this sense, the remarkable diversity and complexity of food provides a constant source of fascination to those involved in its study. Where such fascination occasionally "bubbles over" into the territory of dinner-party conversation, I can only apologize.

Further Reading

Luck, P., and E. A. Foegeding. 2008. "The Role of Copper in Protein Foams." *Food Biophysics* 3:255–260.

McGee, H. J., S. R. Long, and W. R. Briggs. 1984. "Why Whip Egg Whites in Copper Bowls?" *Nature* 308:667–668.

Sagis, L., A. de Groot-Mostert, A. Prins, and E. van der Linden. 2001. "Effect of Copper Ions on the Drainage Stability of Foams Prepared from Egg White." *Colloids and Surfaces A: Physicochemical and Engineering Aspects* 180:163–172.

fifteen

THE MERINGUE CONCEPT AND ITS VARIATIONS

PETER WIERENGA, HELEN HOFSTEDE, ERIK VAN DER LINDEN,

SIDNEY SCHUTTE, AND JONNIE BOER

during a discussion about foams and meringues in our laboratory, two of us (Peter Wierenga and Erik van der Linden) noticed that meringues are usually made with egg whites—quite the revelation for a scientist! We concluded that if we understood foams the way we thought we did, we should be able to make a foam, or derived product, that provides new culinary opportunities. For example, we should be able to make a meringue based on milk alone. At the same time, two others of us (chefs Sidney Schutte and Jonnie Boer) wanted to make a savory meringue. Experience taught us that it is difficult to make and stabilize savory meringue. Therefore, all of us decided to explore the possibility of making milk-based, rather than egg-white, meringue and of testing savory meringue. This was done (in part) by Helen Hofstede.

Here we give an account of the process for developing a savory meringue. We begin with the working hypotheses that led to the design of our culinary experiments. We then describe the experiments and how they enable the creation of savory meringue that is now served at the restaurant De Librije, owned by Jonnie Boer. Our intention is to show that a fundamental knowledge of foams, particularly of regular meringues, can be applied to a new meringue paradigm: savory, not sweet; milk based as opposed to egg-white based. We also provide simple rules of thumb for creating variations on the meringue concept.

The Meringue as Foam

Meringue is typically defined as a mixture of beaten egg whites and sugar. However, different types of meringue exist and each exhibits different properties. The three main types are known as Italian, Swiss, and French. All of them share the same ingredients list, but the method of preparation varies. For a French meringue, the egg whites are beaten at room temperature and the (granulated) sugar is added well into the whipping process. For Italian, the sugar is added as a hot (less than 194°F [90°C]) sugar syrup after the whites are beaten to the medium-peaks stage. For a Swiss meringue, the egg whites are beaten over a bain-marie (double boiler), in which the temperature can be anywhere from 140 to 176°F (60–80°C). Italian and Swiss meringues can be served raw, whereas French meringue is usually baked in an oven (resulting in a solid meringue). Although the preparations are different, all recipes use egg white to form and stabilize the foam.

As discussed in chapters 14 and 16, foams are generally systems in which gas bubbles are dispersed in either one of two phases—liquid or solid. Examples can be found everywhere in daily life: cappuccino, whipped cream, and the heads of beer are all liquid foams. Examples of solid foams include bread, ice cream, and certain pastries, like sponge cake and of course meringue. It goes without saying that the role of gas in any foamed structure is to change the taste, texture, or mouthfeel of a product. For reference, imagine the different eating experiences provided by unwhipped frozen cream and cream that is whipped and frozen at the same time.

FRENCH MERINGUE

2 large egg whites
3½ ounces (100 g) granulated sugar

Preheat the oven to 212°F (100°C). Whip the egg whites either by hand or with a stand mixer. As the whipped substance shows observable standing peaks, add the sugar and continue to whip until stiff. Pipe or spoon to the desired shape and size, and bake until dry (usually between 1 and 2 hours). Let it cool down and enjoy.

We took on solid meringue as a subject of study because the preparation is relatively simple, as seen in a typical recipe for making an egg-based French meringue.

The way in which meringue is made leaves little room for creative variations in taste or structure. While thinking about this problem, we noticed a similarity to ice cream. Ice cream recipes share the same basic ingredients. Still, ice cream is generally produced in many more varieties than meringue. As we set out to create variety in meringues, we dug a little deeper into how ice cream is prepared to obtain its wide-ranging flavors (for more on ice cream, see chapters 17 and 33). How is it there is such a great variety of ice cream flavors that share essentially the same structure, while there is little variation in the appearance and properties of meringue? The answer lies in the understanding of the product. In ice cream, the structural and textural properties are all linked to a standard recipe. To create a specific flavor, we need only add flavor to the mix, which is usually made of cream, milk, and sugar.

We wondered whether the same principle could be applied to meringue. The question then became: Would we be able to use milk as the basis for meringue, and could we borrow the principles from ice cream to create a wide range of meringue products, extending what we learned to savory meringue? We soon realized that it would take more than whipping to get the job done and that physics and chemistry were at the core of the matter.

The Physics and Chemistry of a Meringue

To make a foam, air bubbles need to be trapped in a liquid. These bubbles then need to be stabilized against coalescence, which is understood as the rupture of the liquid films separating the bubbles. There are two ways to achieve stabilization—the first is to gel, or immobilize, the liquid, for instance, with gelatin. The other is to stabilize the surface of the bubbles with proteins. In meringue, the trick seems to be that while the formation of liquid foam is enabled by letting proteins adhere at the bubble interface, it is during the baking step that foam is actually stabilized. This is because as water evaporates during baking, the effective sugar concentration increases, leading to the formation of a solid foam. This foam is stabilized by a layer of sugar that is, for the most part, in a solid, glassy state (chapter 24).

What concentrations of protein and sugar are actually necessary to obtain a nicely baked foam? This can be tested by simply diluting egg white—a 10 percent weight-in-volume (w/v) protein dispersion in water. Surprisingly, even if the egg white is diluted ten times, a stable foam is formed. This illustrates the fact that, in the context of foam formation, there is an excess of protein present in the egg white. Apparently, a concentration of about 1.0 to 1.5 percent is enough to form a stable foam, as illustrated in figure 20, in which the volume of foam is plotted as a function of the concentration of egg-white powder dispersed in 100 mL of water. It seems obvious that increasing the protein concentration beyond 1 percent w/v does not lead to a further increase in foam volume. Bear in mind that in these experiments, the starting volume of the water remained constant.

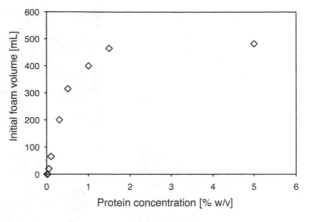

Figure 20 Foam volume at the end of whipping as a function of protein concentration. (Reproduced with permission from Peter Wierenga, Helen Hofstede, Erik van der Linden, Sidney Schutte, and Jonnie Boer, "Variations in Meringues," in *Food Quality, an Issue of Molecule Based Science: Proceedings of Euro Food Chem XIV*, ed. Hervé This and Trygve Eklund [Paris: Société française de chimie, 2007], 1:32–36)

What happens when we hold the absolute amount of protein constant while we increase the amount of water available? This is done by dilution. By increasing the dilution factor, the amount of foam increased. This clearly shows that the final volume of foam obtained from one egg white is directly dependent on the amount of liquid present. Apparently, for all concentrations studied, there is a minimum amount of liquid needed to separate the bubbles from one another. This means that at the point at which the amount of liquid in the system is equal to the amount needed to separate all bubbles, no more air can be incorporated into the foam. This result is analogous to that observed in mayonnaise by both Harold McGee (1990) and Hervé This (2006), who described that more oil could be incorporated into a mayonnaise if the volume of the liquid (as water, vinegar, or lemon juice) was increased—which highlighted the emulsifying power of egg yolk. Therefore, to create more foam, more liquid must be added, making sure the minimum amount of protein is present (1.0–1.5 percent w/v).

The effect of adding sugar to egg-white foam (in the typical amount used for meringue) seems to be similar to that of increasing the volume of the liquid phase. To illustrate this, the microstructure of the liquid foam formed by following the meringue recipe provided, with and without sugar, is shown in figure 21.

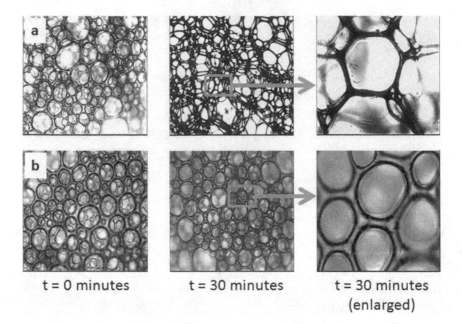

t = 0 minutes   t = 30 minutes   t = 30 minutes
                                                (enlarged)

Figure 21 The 5 percent egg white foams in (*a*) the absence and (*b*) the presence of 200 g of sugar per 100 mL (7 ounces of sugar per 3 fluid ounces) of egg-white solution. (Reproduced with permission from Wierenga et al., "Variations in Meringues")

To understand what happens, recall that in these experiments, the egg white was first beaten to soft peaks and the sugar was then added by gently stirring it into the foam. Of course, with this technique, it is impossible to tell whether all the sugar is being dissolved. What happens upon the addition of the sugar is that the total volume of the liquid phase is increased (figure 22*a*). Because this volume increase happens after beating—that is, no more air is incorporated into the foam—the average distance between the bubbles increases. Looking at this from a different angle, after the addition of the sugar, it is indeed possible to increase the volume of foam by additional whipping. Hence, the way in which the sugar is added, as well as the stage at which it is added, might have an effect on the final foam properties. How these different effects are achieved is not clearly understood and offers a great subject for further study!

Besides increasing the volume of the liquid phase, sugar, in contrast to water, also increases the viscosity. This helps to delay drainage—that is, leaking of liquid from the foam. As evident from figure 22*b*, a significant increase in viscosity is reached only if more than 1¾ ounce (50 g) of sugar is added to 3½ ounces (100 mL) liquid. At such a sugar concentration, the amount of liquid drained from the

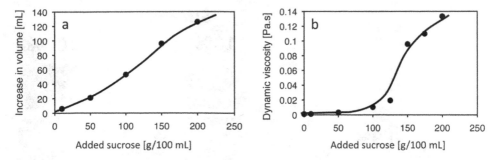

Figure 22 The effect of sugar on (*a*) the total volume and (*b*) the viscosity of the liquid phase (added sugar in g/100 mL H_2O). (Reproduced with permission from Wierenga et al., "Variations in Meringues")

foam after 30 minutes is decreased from 2½ to 2¾ liquid ounces (70–80 mL) to only ¾ to 1 liquid ounce (20–30 mL).

The formation and stability of the liquid foam are not the only important parameters in the making of a meringue—there is the baking and drying provided by the oven. During baking, water evaporation takes place, resulting in an increase in the effective concentration of both the protein and the sugar, which eventually leads to a transition into the glassy state of the system. To develop successful variations of the recipe, it is necessary to understand what factors are important to keeping a voluminous foam during and after baking. As such, foams ranging in concentrations of protein and sugar were baked in an oven at 212°F (100°C) for 2 hours. The resulting structures are shown in figure 23*a*.

When only protein was used, the foam after baking had a rubbery texture, just like cooked egg white. In the absence of sugar, foams made with less than 5 percent egg-white protein collapsed during baking, whereas those made with more protein remained but their textures were elastic and rubbery. As sugar was added, more solid, brittle foams were obtained after baking. Apparently, both protein and sugar can be used to stabilize the foam during baking, but they result in different meringue properties.

By weighing the foam before and after baking, the loss of water can be measured. The results plotted in figure 23*b* show the loss of water versus the initial amount of water present in the sample, for all samples shown in figure 21. Although the texture of the baked foam depends on the ingredient used to provide the solid mass, the amount of water left in the end product does not depend on the ingredients. In other words, water content at the end of baking is the same.

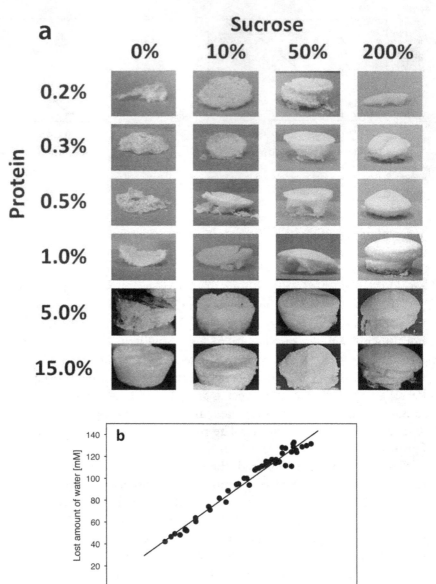

Figure 23 (a) Meringue structure after baking at different protein and sucrose concentrations (% w/v). (b) The water content before baking and the amount of water lost after baking for all samples from figure 23a. (Reproduced with permission from Wierenga et al., "Variations in Meringues")

The results illustrate three important things:

1. At least 5 percent protein is necessary to provide a foam structure that survives the baking step.
2. The solid content of the foam determines the amount of collapse during baking.
3. When sugar is used, a harder, more brittle structure is obtained than when only protein is used.

From Hypotheses to Kitchen Tests to Plates on the Table

The emerging hypothesis is that the meringue recipe can be varied—as long as there is enough protein to form and stabilize the foam, and the amount of foam is maximized to avoid collapse early on in the baking process. The first rule of thumb, therefore, is that for a given amount of liquid, there should be a minimal protein concentration, say 5 percent or more. The second is to make sure that the liquid is maximally whipped; otherwise, the walls of bubbles will become too thick, favoring drainage of liquid and leading to collapse early on during baking.

How do we translate the above to the reality of the kitchen? First of all, even though we used egg-white powder in our experiments, similar results can be obtained with fresh egg whites or other protein sources. A real example is the soy-based macaroon made by Moshik Roth in his restaurant 't Brouwerskolkje—where the planning stages of this book were carried out (see the introduction). Bear in mind that because of differences in protein types and sources, as well as the methods used to produce the powders, differences in taste and functionality are common. If it does not work the first time around, document your findings and keep trying!

To prove that the source of the protein is of minor importance, meringues were prepared using commercially available defatted milk. The protein concentration in defatted milk is around 3.5 percent, which is more than the 1.5 percent needed for foam formation. It was found that nice, stable meringues could be made from defatted milk. The next and obvious step was to dissolve skim milk powder into tomato juice, whip it, and bake it—tomato meringues! At the time, these were served at restaurant De Librije.

A Final Reflection

Our work with meringues has been useful from a gastronomical point of view. We believe we have opened up a range of new applications and ways of thinking about meringue products. Additionally, we have uncovered more about the chemistry and physics of the formation and stability of foams, as well as the relevance of the glassy state in such systems. To our knowledge, this is the first time the concept of meringue has been studied in such a systematic way. From our experimentation, we conclude that a systematic approach to the study of the processes and products in cooking can yield interesting insights for cooks and scientists alike.

Acknowledgment

The authors gratefully acknowledge permission from Hervé This and Trygve Eklund, editors of *Food Quality, an Issue of Molecule Based Science: Proceedings of Euro Food Chem XIV*, to reprint portions of "Variations in Meringues," which appears in that volume.

Further Reading

McGee, Harold. 1990. *The Curious Cook: More Kitchen Science and Lore*. New York: Macmillan.

This, Hervé. 2006. *Molecular Gastronomy: Exploring the Science of Flavor*. Translated by Malcolm DeBevoise. New York: Columbia University Press.

Wierenga, Peter, Helen Hofstede, Erik van der Linden, Sidney Schutte, and Jonnie Boer. 2007. "Variations in Meringues." In *Food Quality, an Issue of Molecule Based Science: Proceedings of Euro Food Chem XIV*, edited by Hervé This and Trygve Eklund, 1:32–36. Paris: Société française de chimie.

sixteen

WHY DOES COLD MILK FOAM BETTER?

Into the Nature of Milk Foam

JULIA MALDONADO-VALDERRAMA, PETER J. WILDE,

AND MARÍA J. GÁLVEZ-RUIZ

have you ever tried to make a real cappuccino with espresso and steamed milk? If you have, you probably know that getting the milk to foam properly is a precise art. In other words, creating the perfect foam for your coffee is not as easy as it sounds.

Foaming milk in a controlled way is essential for creating a genuine cappuccino drink. A proper cappuccino requires a pourable, virtually liquid foam that tastes sweet and rich and stays in the cup. The composition of the milk and the method by which the foam is created will ultimately determine the quality of the foam that is perceived by the consumer. Foaming milk to make a genuine cappuccino is a specialized process. It is generally accepted that one should use cold milk (40°F [4°C]) and rapidly inject steam into the chilled liquid. The steam and rapid mixing form the bubbles that create the foam. The type of milk is important. Most skim and part-skim milks are easier to foam than whole milk; however, foam from whole milk can be more tasty and creamy. In trying to create the perfect foam, we need to consider some important physical and chemical aspects of the process.

Foam Formation

Chapter 14 provides an introduction to the world of foams. Therefore, we know that a foam is generally defined as a mixture of gas bubbles in a liquid. Simply dispersing a gas in water will not form a stable foam, as you have no doubt witnessed. Instead, a very short-lived foam will form—a foam with large bubbles that will rapidly merge, leading to the collapse of the foam. Therefore, air bubbles need a protective layer around them that prevents them from merging (figure 24). This layer can be formed by molecules that actively move toward the surface of the bubbles and stay there. Scientists refer to these molecules as surface-active agents, or surfactants. The surface-active agents in milk are proteins. Three aspects of proteins are important for making and stabilizing the foam:

- Their ability to adhere quickly onto the air–water interface and create a layer around the bubble
- Their ability to remain at the air–water interface
- Their ability to create an elastic network at the interface that enables their flow from the heating vessel to the cup

Bubbles can be made in various ways—for instance, by mechanical agitation (whipping), steam injection, or air bubbling. In all cases, this involves the creation of a surface area between the gas and liquid. To protect the newly created bubbles, proteins adhere to the

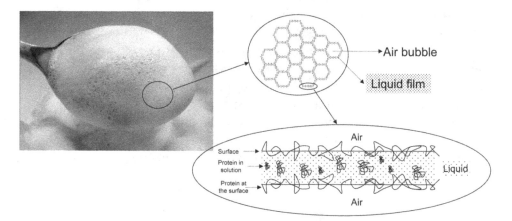

Figure 24 Foam microstructure. The air bubbles are surrounded by liquid films. The liquid films are surrounded by proteins between the air and the liquid, protecting the bubbles.

gas–liquid interface, and the rate of adsorption is a key factor in determining the formation of foam. This rate of adsorption depends on the properties of the proteins, such as their size. Smaller-size proteins can move faster through the liquid onto the created air–water interface, yielding a faster adsorption rate. The ability of the proteins to remain at the interface and to form an elastic network around the bubbles depends on the type and extent of the changes in structure the proteins undergo during whipping.

Apart from protein-related aspects, the solubility of the gas in the liquid phase matters. The more soluble the gas is in a liquid, the more of it is dissolved, and the larger the potential for bubble formation. Thus, foam development is favored by gas solubility.

Foam Stability

Bubbles like those contained in a head of beer or a cappuccino are nothing but air surrounded by liquid. The higher the volume of the foam, for a constant volume of starting liquid, the thinner the layer that surrounds the bubbles. This thinner layer forms because the liquid must spread more to accommodate the extra volume. Because the liquid is heavier than the gas, the liquid drains down through the foam. However, the thicker the liquid, the slower the draining, which helps in the stabilization of the foam (figure 25). As this occurs, the bubbles come even closer together as the thin film between the bubbles becomes even thinner. The thinner the film, the more likely it is to rupture (like a soap-bubble film that spontaneously pops).

Aside from the draining of the liquid, the way the gas flows *within* the foam also affects foam stability. This may sound counterintuitive, but gases tend to move from small bubbles to big ones. As such, small

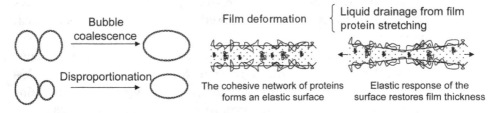

Figure 25 Foam stability. Bubble coalescence implies the rupture of the film separating two bubbles. Disproportionation implies gas diffusion from smaller to larger bubbles. The proteins at the surface protect the bubbles by forming an elastic network. This allows for elastic deformation of the liquid films and prevents rupture.

bubbles become smaller and smaller, and the large ones grow larger and larger. This phenomenon is commonly referred to as disproportionation. Because larger bubbles are more fragile than smaller ones, disproportionation decreases foam stability. Accordingly, gases with low solubility in liquid create more stable foams than those that dissolve easily because the rate at which disproportionation occurs is significantly reduced. An example of this can be found in beer. Foam in a Guinness lasts longer than in other beers. This is because Guinness foam is formed with nitrogen instead of carbon dioxide, and nitrogen is less soluble in water than carbon dioxide.

Surfactants, like proteins, play a critical role in foam stability. As foams are moved around—when a person drinks a beer or scoops foam into a cappuccino, for example—bubbles are squeezed against one another and may fuse. This creates bigger, less stable bubbles. When enough proteins adhere at the bubble interface, an elastic protein network is formed and the susceptibility to fusion is drastically reduced.

It is very important to emphasize that although foam development and foam stability are two different phenomena, they cannot be fully separated. Efficient foaming is achieved only when good stability is obtained, otherwise the bubbles collapse during formation. Hence, the creation of the perfect foam is achieved by compromising these two aspects and finding the right balance.

Application to Milk Foam

Let us apply the concepts described above to the formation and stabilization of milk foams. The usual way to foam milk for a cappuccino is to use a steam wand (like the ones attached to espresso machines). The wand rapidly injects steam into cold milk (40°F [4°C]) until the temperature reaches about 150°F (65°C). Steam injection produces significantly stronger foams compared with bubbling and mechanical agitation.

Milk contains flexible proteins (caseins) and globular proteins (whey). The surface of the bubbles in foam is believed to be composed of a mixture of both types of proteins. Whole milk contains around 3.7 percent fat in the form of milk fat globules, which are in a semisolid state at a temperature of less than 41°F (5°C). However, at a higher temperature, which is reached when steam is injected into the milk, the fat will melt. Liquid fat spreads more easily over the surface of the nascent bubbles and inhibits foam formation—just as a bit of egg-yolk infiltration inhibits the whipping of egg whites. This effect is particularly important during the early stages of foam development because the proteins have not yet formed a strong, elastic layer on the bubble surface. Conversely, the solid fat globules of cold whole milk will not inhibit foam formation. It goes without saying that skim milk is more suitable for foam making; however, some say it also makes less creamy foams.

The solubility of air in liquid decreases with increased temperature, so steam injection offers a good compromise between foam formation and stability. Air is more soluble in cold milk and this favors the formation of foam. Then, as the foam is heated via steam injection, and the solubility of the air in the liquid decreases, proteins are mildly heated, which favors their mobility on the bubble surface. As the milk is heated further, proteins open up, or denature, and start to associate with one another, creating an elastic protein network on the surface of the bubbles. This significantly increases the stability of the foam.

However, hot milk can have a negative effect on many aspects of both foam formation and stability. When hot, milk becomes less viscous and therefore the liquid drains faster from within the bubbles. Second, the solubility of air in hot milk decreases, inhibiting foam formation. Last but not least, and relevant only for whole or part-skim milk, fat globules are liquid and able to cause film rupture as already described.

All this means that, at least in principle, heated and then cooled milk should be suitable for foaming. However, milk that has been ultra-high temperature treated (that is, sterilized) produces weaker foams, presumably because the extremely high processing temperatures (268°F [131°C]) have reduced the ability of proteins to adsorb to and stabilize the bubbles.

A firm grasp of the science of foams yields valuable tips for the creation of the perfect cappuccino drink. Foams produced with skim milk are more stable than those made with whole milk because of the detrimental effect of fat on the liquid foam. If you must use whole milk, make sure it is cold. Also, it is recommended that a stainless-steel jug be used to create the foam. Steel will dissipate some of the heat, allowing more time for air to become infused into the cold milk before it gets too hot and fat globules become liquid.

Given the complexity of the foaming process, the different varieties of milk on the market, and the array of cappuccino preparation methods, many questions about the foaming process remain unanswered.

However, we believe that understanding some basic concepts behind the physics and chemistry of foam formation and stabilization can enable espresso lovers to take their favorite beverage to a whole new level.

Foam, anyone?

Further Reading

Arboleya, J.-C., I. Olabarrieta, A. Luis-Aduriz, D. Lasa, J. Vergara, E. Sanmartín, L. Iturriaga, A. Duch, and A. Martínez de Marañón. 2008. "From the Chef's Mind to the Dish: How Scientific Approaches Facilitate the Creative Process." *Food Biophysics* 3:261–268.

Borcherding, K., P. Chr. Lorenzen, W. Hoffmann, and K. Schrader. 2008. "Effect of Foaming Temperature and Varying Time/Temperature-Conditions of Pre-heating on the Foaming Properties of Skimmed Milk." *International Dairy Journal* 18:349–358.

Goh, J., O. Kravchuk, and H. C. Deeth. 2009. "Comparison of Mechanical Agitation, Steam Injection, and Air Bubbling for Foaming of Milk of Different Kinds." *Milchwissenschaft* 64:121–124.

Kamath, S., T. Huppertz, A. V. Houlihan, and H. A. Deeth. 2008. "The Influence of Temperature on the Foaming of Milk." *International Dairy Journal* 18:994–1002.

seventeen

ICE CREAM UNLIMITED

The Possibilities of Ingredient Pairing

ELKE SCHOLTEN AND MIRIAM PETERS

ice cream is a popular frozen dessert consumed the world over. It is typically enjoyed as a cool warm-weather treat, and a range of varieties can be found across many cultures. Although everyone loves ice cream, Australians and New Zealanders, who eat the frozen treat all year long, seem to have the largest annual per capita consumption, with 18 quarts (17 L) and 17 quarts (16 L), respectively.

Ice cream is essentially milk, water, cream, and sugar. But change the ingredients slightly and standard ice cream can become gelato (custard-based ice cream), sorbet (nondairy, fruit-based frozen dessert), frozen yogurt, or fruit ice. Ice cream is popular because of its mouthfeel, conveying a sensation of decadence on the tongue. Just out of the freezer, it is cold and hard, and as it melts, it turns into a smooth, creamy liquid. The combination of flavor, texture, and the cooling sensation in the mouth will determine whether we like it. Although most ice creams have similar ingredients, the mouthfeel and the sensations they impart are different. Sorbets, for example, normally feel a bit colder than ice creams, and their texture is completely different. Some ice creams melt very easily and disappear very quickly on the tongue, while others last. Why are there so many differences, and where do these differences come from? Why do all ice creams contain sugar? Is it possible to make ice creams without sugar?

Looking closer at ice cream and sorbet, we can recognize that these frozen products are complex systems whose microstructure can be described as ice crystals and air bubbles embedded in an unfrozen sugar solution. The unfrozen sugar solution is often referred to as the freeze-concentrated matrix, as upon freezing, water is removed from the solution as ice, effectively increasing the sugar concentration in the unfrozen phase. The structure of ice cream—whether it is gelato or sorbet—is determined by the proportion of these different elements (ice, air, and the unfrozen matrix) and is dependent on the specific ingredients and the process by which the ice cream is produced. The microstructure of the ice cream is important because it determines the dessert's sensorial properties, such as hardness, coldness, rate of melting, creaminess, and fluffiness. It thus determines how we perceive ice cream when we eat it, the ease with which the ice cream can be scooped from its container, and how long the ice cream can be stored without losing its characteristic texture.

An understanding of how all these elements are interrelated can help us manipulate ice cream's properties to achieve certain effects.

The Ice Element

The most important element in ice cream is the ice—solid crystalline water. Liquid water becomes solid at 32°F (0°C). Ice cream is normally consumed at a temperature between 7 and 14°F (−14 and −10°C). If ice cream consisted only of water, all the water would turn into ice at these temperatures, and the "ice cream" would be very hard and almost impossible to eat. Therefore, the amount of ice has to be controlled; that is, part of the water needs to remain in the liquid state. For the water phase to be only partially frozen, the freezing point of water has to be lowered. Adding different ingredients, such as sugars, salts, alcohol, and other additives, will aid in the lowering of the freezing point. These ingredients can be considered (for our purposes) "antifreeze" agents, and as they lower the temperature at which the water starts to freeze, they will determine the amount of solid ice present and, as a result, the hardness of the ice cream.

Freezing Point Depression

As already described, when small molecules, such as sugar, salt, or alcohol, are added to water, the temperature at which the water starts to freeze will be less than 32°F (0°C). This effect is known as freezing point depression. By adding these different molecules to water, the water molecules cannot easily congregate to form a crystal. The more nonwater molecules present, the more difficult it is for the water to become a solid, and the more the solution requires cooling to create ice (solid crystalline water). The *number of molecules* therefore deter-

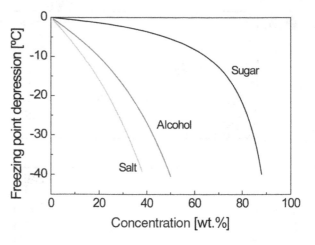

Figure 26 The freezing point depression for different "antifreeze" agents as a function of their concentration.

mines the freezing point depression. On a per-weight basis, molecules of lower molecular weight (such as salt) are relatively more effective in lowering the freezing point than large molecules (such as sugar); 1 gram of salt contains about 4 times more molecules than 1 gram of sugar. Figure 26 shows the relative freezing point depression for different amounts of salt, sugar, and alcohol. For sugar to register a *similar* freezing point depression as salt or alcohol, more of it is required.

Because salt and alcohol can also lower the freezing point, it is possible to make a sugar-free ice cream by using specific amounts of other antifreeze agents. This can be seen in figure 27, which shows the amount of ice present as a function of temperature for different additive concentrations. The dashed line represents the addition of a higher concentration of additive compared with the solid line. When the solution contains 10½ ounces of sugar per quart (300 g/L), we can see that at 5°F (−15°C), around 60 percent of the water is ice, and when the temperature rises to 23°F (−5°C), the amount of solid ice is about 40 percent. In the case of salt as a replacement for sugar, 2½ ounces (70 g) are enough to give the same effect as 10½ ounces (300 g) of sugar at 5°F (−15°C) (about 60 percent of the water is ice). However, when the temperature of this salty "ice cream" is increased to 23°F (−5°C), we see that all the ice has melted, whereas in the case of the sugar solution, ice is still present at 23°F (−5°C). The figure shows essentially the same behavior for alcohol, but with the freezing lines again somewhat differently positioned. Therefore, even though sugar can be replaced by other additives to make ice cream,

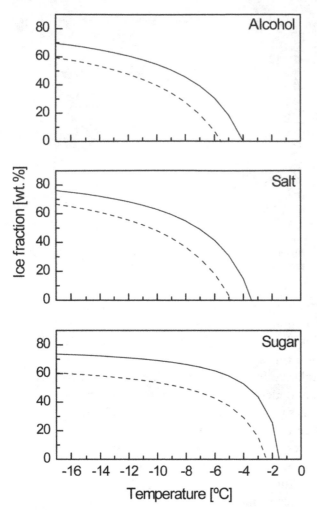

the melting behavior of the ice cream will be different—keep this in mind!

As these antifreeze agents affect the melting behavior, it is reasonable to believe that they will also have a significant impact on the ice cream's sensorial properties, such as hardness and coldness. To create an ice cream with a pleasant mouthfeel, the relative amount of antifreeze agent that can be added is limited. For example, too much sugar and the ice cream becomes too soft (and too sweet); however, too little sugar and the ice cream is too hard and not sweet enough. A delicate balance therefore must be found.

Figure 27 The ice fraction as a function of temperature for different "antifreeze" agents. For alcohol, the solid line refers to 90 g and the dotted line to 120 g. For salt, the solid line represents 50 g and the dotted line 70 g. The amount of sugar added is 200 g (solid line) and 300 g (dotted line).

Freezing Temperature

Solero Shots and Calippo Shots are little ice-sphere treats found in European supermarkets. These ice spheres are a few millimeters in diameter and instantly melt in the mouth, leaving behind an intense cooling sensation. The temperature at which the shots are eaten is the same as regular ice cream. So why do these ice spheres melt so fast and feel so cold?

It is not only the presence of ice but the size of the ice crystals that influences the perception of coldness. A few very large ice crystals can have the same volume as a lot of very small crystals. When you eat ice cream that has large crystals, it will have a gritty and unpleasant mouthfeel (this is otherwise desirable in granita). Ice creams should therefore have small crystals. The size of the ice crystals can be controlled by the temperature of the ice cream maker. The colder the freezing barrel, the more likely new (small) ice crystals will be formed. Conversely, when the ice cream maker is not cold enough, far fewer small ice crystals are formed, and therefore the growth of large ice crystals is favored. As a result, fewer but bigger ice crystals develop.

It follows that the surface area of the small ice crystals is much larger than that of large ice crystals, even when the total amount of ice is the same. The temperature of the ice cream increases as it is consumed and the larger total surface area of the small ice crystals encourages melting. During the melting of the ice, the ice cream will absorb heat from the mouth, which is perceived on the human tongue as a cold sensation. The faster the ice cream melts, the more heat it absorbs, and thus the colder it feels. So, in eating ice cream, it is not only the amount of ice and the ice cream temperature that determine the sensation of coldness but the manner in which the ice cream was made.

This means you can use the temperature at which you make the ice cream to influence the melting of the ice and thereby the sensation of coldness. The lower the temperature during production, the colder the ice cream will feel. The Solero Shots and Calippo Shots ice spheres are made by dipping the ice cream premix in liquid nitrogen, which is very cold (–320°F [–196°C]), and therefore it will form a large number of very tiny ice crystals. So, even though the temperature at which you eat small ice spheres is the same as any other commercial ice cream, they feel much colder because of the process by which they were made. The lucky few who have dined at high-end restaurants that practice science-based cooking have perhaps experienced the gastronomic use of liquid nitrogen. This results in ice cream with very small ice crystals and therefore a high melting rate—and strong cooling sensation.

The Air Element

During the freezing of ice cream, air is mixed in, forming air bubbles. For the air bubbles to stay incorporated inside the mixture, they have to be stabilized (like foam in a cappuccino). For this to happen, ice cream needs ingredients that have a preference for literally sitting on the air-bubble interface and preventing the air bubbles from fusing together. Ingredients that have this ability include proteins and partially solid fats, both of which are present in dairy products, such as cream and milk. Ice cream that contains milk and cream thus contains more air bubbles than sorbet, which is made without them. This has a major impact on the texture of the ice cream: as one can imagine, the more air bubbles present, the fluffier and softer the ice cream. This impact on ice cream texture is enhanced by an additional, less trivial effect: the air bubbles that are incorporated into the ice cream also prevent ice crystals from forming a large solid network. The air bubbles thus effectively stabilize the ice crystals and help preserve a smooth ice cream texture during storage.

The Matrix Element

We have seen that the ice element and the air element are mainly responsible for the sensorial properties, such as hardness, coldness, and fluffiness. The unfrozen matrix contains the bulk of the dessert's ingredients, apart from ice and air. It contains sugar but possibly also salt and fresh fruit or fruit concentrates. As the system freezes, the liquid phase gets more concentrated because freezing removes water in the form of ice. The concentration of these additives will determine the viscosity of the aqueous phase and therefore partly determine the texture of the ice cream. For instance, sugar is known to have a positive effect on ice cream structure and texture. Concentrated sugar solutions are viscous—at low temperatures, in particular—because sugar has the ability to incorporate air (similar to fat and protein) and inhibit the mobility of the air cells in the ice cream matrix. Therefore, sugar lends a smooth texture and renders the ice cream more stable. In addition, it lends body to the matrix, which remains when the ice cream is molten. Other ingredients that make the matrix more viscous are pectin (naturally present in fruit) and gelatin. However, these compounds, also known as biopolymers, have very large molecules, so-called biopolymers, and while impacting the viscosity of the matrix, they do not influence the freezing behavior nearly as effectively as sugar does.

Ice Cream Versus Sorbet

As we have seen, the ice, air, and matrix determine the mouthfeel of the ice cream and, to a great extent, the stability of the ice cream. Figure 28 shows a simple schematic of the microstructure of ice cream. The fat particles and the proteins are situated at the interface between the matrix and the air bubbles. In its frozen state, between its consumption temperatures of 7 and 14°F (−14 and −10°C), ice cream normally consists of about 50 percent air, 30 percent ice crystals, 15 percent matrix, and 5 percent fat. With this particular composition, about 60 percent of the water is present as ice.

The main difference between ice cream and sorbet is the air element. Ice cream always contains dairy products, while sorbet is largely a sugar solution. As discussed, milk fat and proteins provide the stabilizing effect for the incorporation of air into ice cream. Sorbet, which lacks these bubble-stabilizing ingredients, is consequently less creamy and contains much less air than ice cream. To increase the overall stability of sorbet during storage, gelatin, which is a protein, can be added. As gelatin consists of molecules that are much

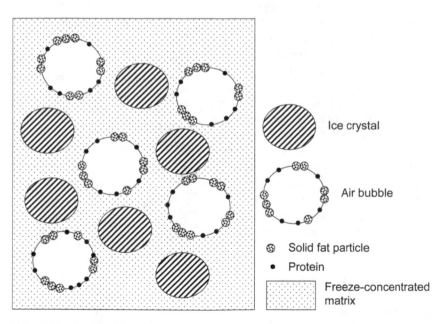

Figure 28 A schematic of the microstructure of ice cream.

Ice crystal

Air bubble

⊛ Solid fat particle

• Protein

Freeze-concentrated matrix

larger than those in sugar and salt, it does not significantly affect the freezing point depression because this property is mainly influenced by small molecules. However, gelatin seems to decrease ice crystal growth, given its ability to bind with water.

Ingredients of Ice Cream and Sorbet

As has been noted, there are only a few key elements to making ice cream or sorbet:

- Water (liquid and solid [ice])
- "Antifreeze" agent—that is, an agent that lowers the freezing point and reduces the amount of water that is frozen at any given temperature
- Stabilizer (to stabilize the air bubbles)
- Fat (in the case of ice cream)

This means that almost anything can be made into an ice cream, as long as all the basic ingredients are naturally present at a specific ratio. Water can be found in any type of food or beverage, such as juices, sodas, fruits, vegetables, and beer. When these liquids are cooled, part of the water will change into solid ice, depending on the temperature of the ice cream maker and the additives present in the liquid. In order to control the amount of ice, a specific proportion of freezing point–lowering agents should be present. This provides a creative oppor-

CHERRY-KRIEK SORBET

You can easily make sorbet or ice cream with fresh fruit at home. Fresh fruit already contains a certain amount of sugar and therefore the added sugar will be different for different fruits. Strawberries contain much more sugar than lemons, so more sugar should be added to lemon ice cream. Alcohol can also be used in combination with fresh fruit. However, alcohol provides an antifreeze effect because it dramatically lowers the freezing point, so in its presence, the amount of sugar should be limited.* We have chosen to make a cherry-kriek sorbet. Kriek is a special style of Belgian beer that is made from morello cherries. To enhance the cherry flavor of the beer, we have added cherries. The beer provides the antifreeze agents alcohol and sugar and the cherries provide the sugar. The sugar and the thickeners (pectin) in cherries provide viscosity to the finished product.

23 ounces (650 g) bottled pitted sweet cherries
10 ounces (300 mL) kriek
1¾ ounces (50 g) granulated sugar

tunity in the amount and type of antifreeze agents that are selected (while keeping in mind how these affect flavor). Instead of a sweet ice cream, a salty or an alcoholic "ice cream" can be made. However, remember that when replacing sugar with other ingredients, the texture might change; without the sugar, the matrix will be less viscous. However, this also provides an opportunity for playing with a fifth building block: a texturizer, such as gelatin, can be added to compensate for the loss of texture. As already mentioned, protein and fat are often added in the form of cream and milk. However, any other product that contains protein or fat can be added, such as cheese, yogurt, milk powder, fish, meat, peanut butter, or chocolate. The number and combinations of ingredients are endless, and the only limit is your own imagination.

Grind the cherries in a blender and add the kriek and sugar while continuing to blend. Allow the sorbet mix to cool down in the fridge for an hour. Place your ice cream maker's freezer bowl into the machine. The bowl should have been stored overnight in the freezer. Pour the sorbet mix into the ice cream maker and start the machine. The ice cream maker cools and whips the sorbet mix simultaneously to incorporate the air and to allow ice formation. Remove the sorbet after about 45 minutes. As the temperature of the sorbet is still high (about 21°F [−6°C]) and the texture very soft, let the ice cream harden in the freezer for a few hours.

* For this recipe, the freezing point depression is determined by the proportions of sugar and alcohol. The total amount of sugar in the ice cream mix is 5 ounces (145 g) (20 percent coming from 23 ounces [650 g] of cherries and 5 percent from 10 ounces [300 mL] of beer). This gives a freezing point depression of −1.4°F (−0.817°C). The alcohol in the beer (½ ounce [15 g]) leads to an additional −1.1°F (−0.617°C). In total, this gives a freezing point depression of 2.5°F (which corresponds to a freezing temperature of 29.5°F [−1.4°C]). However, as the melting curve of ice with alcohol is different from that of sugar, the sorbet will melt faster than regular sorbet. It will therefore provide a different mouthfeel.

Innovative Sorbet and Ice Cream Recipes

Beyond the basic building blocks of liquid, antifreeze, stabilizer, and fat, it is possible to make ice cream with any flavor combination. We give two recipe examples—one sorbet and the other ice cream—both of which can be made with a small home ice cream maker. The recipes originate from the course Advanced Molecular Gastronomy. It is taught at Wageningen University, in Wageningen, the Netherlands. Students design their own ice cream, taking into account ingredient functionality. The recipes presented here can be optimized according

TZATZIKI ICE CREAM

In savory ice creams and sorbets, sugar is not as desirable because it can make the product too sweet. In this case, sugar can be replaced with a bit of salt. A little goes a long way in frozen desserts, so proceed with caution. And because so little of this antifreeze ingredient can be used, it is sometimes not possible to achieve the targeted freezing point depression with salt alone.* To compensate for this, gelatin can be added to control the crystal growth and therefore the texture. The gelatin also helps stabilize the incorporated air bubbles. In this tzatziki ice cream recipe, we have used cucumbers in combination with Greek-style yogurt, which provides additional proteins to stabilize the air bubbles. In its unfrozen form, tzatziki is a Greek appetizer often served with bread; it is also used as a sauce for souvlaki and gyros.

16 ounces (450 g) cucumber
2 g powdered gelatin
17½ ounces (500 g) Greek-style yogurt
1¾ ounces (50 g) granulated sugar
½ ounce (15 g) salt
Add ground garlic and pepper to taste

Grind the cucumber in a blender and add the powdered gelatin while stirring gently. After the gelatin has dissolved, add and blend together the yogurt, sugar, and salt. Add garlic and pepper to taste (note that the frozen product will have a less intense flavor). To speed up the freezing process, the ice cream mix should be cooled in the refrigerator for approximately 30 to 45 minutes before being poured into an ice cream maker.

This ice cream can be made at home with a small ice cream maker in which a freezer bowl is used to control the temperature. This bowl is stored in the freezer until the correct temperature is reached. As the freezing temperature determines the size of the ice crystals, the bowl temperature should be as low as possible. Churn the ice cream mix in the machine for about 45 minutes, after which the ice cream should harden in the freezer for a few hours. This ice cream is intended

to personal preference; when the ice cream is too hard, add a little bit more antifreeze agent. When it is too soft, take some out. To incorporate more air, add more proteins or gelatin.

A homemade ice cream (or sorbet) usually contains between 18 and 26 percent sugar, 7 and 15 percent fat, and 4 and 8 percent protein. The freezing point for an ice cream with 26 percent sugar is about 29.5°F (−1.4°C). Thus, for an ice cream with different antifreeze agents, the sum of the freezing point depression of the individual ingredients should be around 2.5°F.

These recipes show that if you have a basic understanding of how the microstructure of frozen desserts is organized, you can easily manipulate it through changes in the ingredients. With some imagination, anyone can create the most interesting ice cream—one that can be served in the no longer distant future.

Further Reading

Goff, H. D. 1997. "Colloidal Aspects of Ice Cream: A

Review." *International Dairy Journal* 7:363–373.

——. 2002. "Formation and Stabilization in Ice-Cream and Related Products." *Current Opinion in Colloid and Interface Science* 7:432–437.

Pei, E. X., Z. J. Schmidt, and K. A. Schmidt. 2010. "Ice Cream: Foam Formation and Stabilization—A Review." *Food Reviews International* 26:122–137.

to be eaten immediately after it is made. Upon storage, the texture will change because the ice crystals will continue to grow; in fact, the ice cream's overall texture will become progressively more icy during storage. In commercial ice cream of this type, additional stabilizers are often added to extend the shelf life. Therefore, do not wait too long to eat your frozen tzatziki!

* In this recipe, the sugar (1¾ ounces per quart [50 g/L]) provides a freezing point depression of –0.6°F (–0.3°C). The salt provides an additional freezing point depression of –1.4°F (–0.8°C), which leads to a total freezing point depression of –2°F (that is, the system freezes at 30°F [–1.1°C]). As the freezing point depression is lower than that for sweet ice cream, gelatin is added to control the texture.

eighteen

EGG YOLK

A Library of Textures

CÉSAR VEGA

according to some, egg yolk is the best sauce in the world—I could not agree more. Who does not love the silky, salty, warm taste of a runny egg yolk? More people than I would have thought, regrettably (for them). What are the reasons behind the polarizing attitudes toward runny egg yolk? Texture? Taste? Food safety? Ignorance?

Runny yolks find their way onto our tables in one of three ways: as poached, soft-boiled, or sunny-side-up eggs. In these preparations, the culinary objective (or more correctly, *my* objective) is to cook the white long enough that it sets, while warming up the yolk to a point that it develops some viscosity, or a thickened mouthfeel, but remains free to flow. Getting the right texture is easier said than done. At least in the case of a poached or fried egg, we can see the color and texture developing before our eyes, but what about a soft-boiled egg? There is no way to know until it is already—hopefully not—too late.

Definitions

Before we jump into deeper deliberations, let me first reflect on the term *soft-boiled*. What does it describe—the cooking conditions or the desired texture? Cooking in boiling water implies that the cooking temperature is about 212°F (100°C). It follows that "soft" would refer to the texture of the egg, but it is still unclear to what part of the egg—the yolk or the white? To complicate things even more, some may argue that "soft boiling" is a synonym for *simmering*, also known as the "lazy bubble stage," in which water temperature can be anywhere between 180 and 195°F (80–90°C). I hereby propose that when we refer to shell-on eggs cooked by immersion in hot water, we are speaking of the desired texture of the *yolk*. This is because, as I will demonstrate, any yolk texture is achievable by selecting the right time–temperature combination.

But what is an egg yolk, exactly?

According to Harold McGee (2004), the word *yolk* comes from the Old English word *geolu* (yellow), which in turn, derives from an Indo-European root meaning "to gleam or glimmer." Technically speaking, raw egg yolk is simply an emulsion of fat in water. It represents approximately 32 percent of the weight of an intact egg and it is composed of about 50 percent water, 32 percent fat, and 17 percent protein. Much of its color, an element of quality, comes from a family of compounds known as carotenoids, which are normally absorbed from the animal's corn or grass feed. Many of the proteins and lipids contained in an egg yolk provide much of its known functionality as a foaming, emulsifying, or texturizing agent. It is then not surprising to find it as part of a wide range of dishes, savory and sweet alike, in which the cook takes advantage of one, two, or all three functionalities in a single dish. Its emulsifying power is obvious in the making of mayonnaise, hollandaise, and gribiche sauces. Crème anglaise and crème brûlée are possible thanks to the egg yolk's water-holding capacity, whereas sabayon relies on the yolk's thickening and foaming properties. Although I have a deep personal interest in all these functions, I thought that I should begin with the thickening process that the yolk undergoes when eggs are simply immersed in hot water.

The Gallinaceous Enterprise

There are countless ways of preparing soft-cooked eggs, probably as many as there are home cooks. What is your way? Mine is as follows: I boil water; add the eggs directly from the refrigerator (up to five eggs at once, provided there is enough water to immerse the eggs by at least ¼ inch [0.5 cm]); keep them in boiling water for about 6 minutes; remove them and cool them under cold running water—

the usual and rather gratifying result is a set, soft, and pliable white married to a yolk that is almost as viscous as honey. Regardless of the method, what matters is the outcome: your desired outcome.

Some professional cooks have taken a relatively new approach to the cooking of eggs. They realize that cooking eggs in boiling water allows for little or no control of the speed at which the eggs are cooked. The use of a temperature-controlled water circulator, or water bath, allows for new possibilities. Using this device, chefs have now ventured into cooking eggs at relatively low temperatures (140–160°F [60–70°C]) for relatively long periods of time (at least 1 hour). As a result, it is rather common to find the so-called $6X$°C egg on many menus around the world. I call it the $6X$°C egg because the second digit varies, depending on the restaurant (from 1 to 5, but a 7 can be found as well). To make matters more complicated, the cooking time for the $6X$°C egg also varies. It might seem reasonable to assume that the texture of the eggs cooked under these rather variable conditions should be different. But is it?

Some scientists and chefs familiar with this technique suggest that once the desired internal temperature of the yolk has been reached, the egg can remain at the temperature for a long time, its texture will not be influenced. I found this hard to believe and, to some degree, an oversimplification of the physical chemistry involved: egg cooking is all about texture development, a process driven by protein denaturation and the rate at which it occurs. In this context, the time element is equally as important as the temperature element. Everybody knows that an egg cooked in boiling water for 5 minutes tastes and feels completely different from one cooked for 15 minutes! Several questions are key to my gallinaceous enterprise: Are there *significant* textural differences among all these $6X$°C eggs? How can the texture of these eggs be better described? How does the chef conclude which time–temperature relationship is optimal? How versatile is the technique (that is, can we get the same results by using different time–temperature combinations)? Only a systematic and experimental approach can shed light on the matter.

To learn a bit more, I called a few chefs to find out under what conditions they cooked their $6X$°C eggs, and this is what I found: Chef 1 cooked the eggs at 62°C for 2 hours; Chef 2 at 62.5°C for 1 hour; Chef 3 at 61°C for 1 minute per gram of egg; Chef 4 at 63.5°C for 55 minutes; Chef 5 at 63°C for 1 minute per gram of egg; and Chef 6 at 62.5°C for 40 minutes. How different was the texture of the eggs cooked under these different conditions? I had to find out.

I decided to study the cooking of the yolk and white separately because their composition and location within the egg pose different practical challenges. My focus—based on my proposal that the texture of egg yolk indicates the degree of doneness of the egg—was then on the texture development, or thickening, of the yolk. To grasp the effect of temperature and time on egg yolk thickening, I first needed to measure the increase in temperature of the yolk once the egg was placed inside the water bath. After all, the thickening process is primarily dependent on temperature. I found that the temperature increased linearly with time until it reached about 120°F (50°C), after which it started to slow down as it approached thermal equilibrium. Knowing the "heating rate" of the yolk enables the mimicking of such conditions in the texture analyzer and accounts for the overall cooking time. The experiments were conducted using eggs weighing 2½ ounces (70 g).[1]

Avoiding oversimplification, I divided the heating process of the yolk into four distinct steps. Step 1 is the time it takes for the yolk, at a fixed water bath temperature (for example, 147°F [64°C]), to reach an internal temperature of 140°F (60°C), which is the point at which the most labile protein in egg white, ovotransferrin, denatures.[2] Step 2 is the elapsed time between 60 and $6X$°C (where X is the second digit of the desired cooking temperature, in this case 4), in which nonnegligible changes in the overall egg texture occur. This is because the heating rate is now considerably slower, and 10 minutes or more can elapse before the desired temperature is reached. Step 3, the isothermal step, corresponds to the time the egg is cooked at $6X$°C. Step 4 is the time it takes to cool the egg, in ice water, back to room temperature, which stops the cooking process (and thereby the protein denaturation process).

Recall that the aim of my gallinaceous enterprise was to be able to quantify and describe the textural changes of egg yolk as a function of time and temperature. This meant I had to evaluate how sensitive egg yolk is to small changes in either cooking time or temperature. In this context, is there a significant *textural* difference in yolks cooked for 15 or 30 minutes at 149°F (65°C) or by holding them for 30 minutes at 145 and 147°F (63 and 64°C)? To answer this question, I tested a total of sixty-six different time–temperature combinations.

1 Significant differences can be expected when using eggs that are at least 5 grams more or less heavy than the ones used in this experiment.

2 Even though the focus of the study was on egg yolk, it was important to include the temperature sensitivity of all proteins because eventually they will be integrated into one "whole egg" scenario.

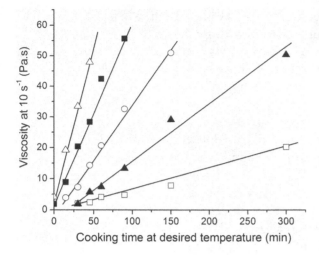

Figure 29 Viscosity (at 10 s^{-1} and 23°C) of cooked egg yolks as a function of isothermal cooking time at different temperatures: 60 (□), 62 (▲), 63 (○), 64 (■), and 65°C (△). The lines represent the fit of the linear regression of the data.

I was so impressed by the results! It seemed that, no matter what the temperature the egg was cooked at, it continued thickening *linearly* with time (figure 29). This is the same as saying that for every minute spent at 6X°C, the viscosity increased by the same amount, which suggests that we might actually be able to predict or, better yet, design the final texture of an egg yolk!

What was equally fascinating was how an increase of only 1.8°F (1°C) in cooking temperature made a tremendous difference in the final texture of the yolk. Figure 29 shows the final viscosity of egg yolks cooked at different and constant temperatures (each set of points) for different lengths of time. You can observe that, for instance, cooking an egg for 30 minutes at 140 or 144°F (60 or 62°C) makes absolutely no difference in the yolk's final texture (2 Pa.s). (For reference, a viscosity of 1 Pa.s is already one thousand times more viscous or thicker than water.) The relative and *physical* meaning of these numbers will become clearer later. If the temperature is raised by just 1.8°F (1°C) to 146°F (63°C), the viscosity increases by almost a factor of four (7 Pa.s). Take the temperature up a notch to 147°F (64°C), and the viscosity shoots to 20 Pa.s; at 149°F (65°C), it is 33, and at 151°F (66°C), it is 54 Pa.s (not shown). I also cooked the eggs at 152, 154, 156, and 158°F (67, 68, 69, and 70°C), but the yolks were way too viscous for the measurement to have any practical meaning. I was now convinced that something very useful would come out of this.

What do all these Pa.s units really mean to you, me, or the chef, anyway? How can we translate these changes in viscosity into a language we all can understand and apply during cooking? There is a need, at least in my mind, for descriptors of texture for cooked egg yolk. Terms like *viscous*, *runny*, or *thick* remain inadequate and ambiguous for the wide range of textures that can be achieved. Perhaps

Table 1 Viscosity (at 10 s⁻¹) of a Series of Thick Edible Fluids at Room Temperature

Food	Viscosity in Pa.s (at 10 s⁻¹)
Whipping cream	0.02
Raw egg yolk	0.09
Pancake syrup	0.96
Chocolate syrup	1.4
Sour cream (17% fat)	2.9
Greek-style yogurt	3.0
Molasses	3.3
Sweetened condensed milk	6.8
Mayonnaise	12.1
Ready-to-eat chocolate pudding	13.8
Honey	18.3
Nutella	28.1
Cookie icing (fresh)	29.3
Toothpaste	43.8
Marmite	43.9

Source: C. Vega and R. Mercadé-Prieto, "Culinary Biophysics: On the Nature of the 6X°C Egg," *Food Biophysics* 6 (2011): 152–159.

the key in refining our description of texture is to associate yolks with other food products, making it possible to describe—and even better, demand—a sought-after eating experience. So, what other foods does thickened yolk look or feel like? Table 1 lists the viscosity of a variety of foods that are often perceived as thick or viscous. The values were measured at conditions that resemble the deformation that foods undergo within the mouth. The significance of the table is that it represents our own mini library of textures that can be used as a tool to describe the texture of cooked egg yolk. The library spans from relatively fluid foods, like pancake syrup, to the more solid toothpaste and Marmite (the British yeast spread).

So, what is the texture you look for when cooking, or more important, eating egg yolks? We now have a series of model viscous foods from which to choose. Say you like egg yolk with a honey-like texture. What does this mean in terms of time–temperature settings? From the texture library, we can observe that this corresponds to a viscosity of roughly 18 Pa.s. Next, you would need to plan ahead and decide whether you have time to spare (that is, at least 1 hour) or whether you are cooking "à la minute," which means you would need to plate the dish within the next 10 to 20 minutes. If you were to use figure 30 to find out how long you would need to cook your eggs to achieve a honey-like texture, you would either hold them for about 300 minutes at 140°F (60°C), 60 minutes at 146°F (63°C), or 15 minutes

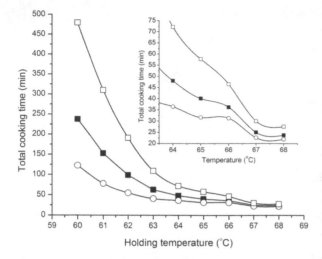

at 149°F (65°C). Now, if you stop and reflect for a moment (but please, not while the eggs are in the pot!), you might notice that figure 29 plots the isothermal time only. Therefore, it does not include the time it takes for the eggs to reach the desired cooking temperature. As it happens, I have created a master plot that not only accounts for this but also eliminates the need for dealing with rheological units (figure 30).

Once the chef or home cook is clear on the sought-after egg yolk texture, or mouthfeel, this plot be-

Figure 30 Isoviscosity lines that relate the holding temperature to the total cooking time needed to develop a characteristic texture in egg yolks (by immersion of eggs in a water bath). Each line represents a desired texture—any combination of time and temperature that falls on or close to any line would render the texture assigned to that line: (O) sweetened condensed milk, (■) chocolate pudding, and (□) cookie icing. The inset shows a close-up of the high-temperature region, where cooking occurs at a faster rate.

Figure 31 Visual cue into the viscosity of egg yolks cooked in the rheometer at a constant temperature of 63°C. The cooking times from left to right are 15, 45, and 90 minutes, respectively. The photos were taken once the samples were cooled to room temperature and 15 seconds after removing the bottom cap of the sample holder (held horizontally).

comes the tool to define what the optimal cooking conditions will be. If we are looking for a thick yolk, like cookie icing, we would need to cook the egg for either 8 hours at 140°F (60°C), a bit more than 5 hours at 142°F (61°C)—3 hours saved by only 1.8°F (1°C)!—or about 45 minutes at 151°F (66°C). The lower the cooking temperature, the more forgiving the process becomes. The corollary is that cooking at 153°F (67°C) or higher and having the desire to create significantly different yolk textures will require an expert familiarity with the technique (see inset in figure 30). To visualize how the structure, or viscosity, and hence the mouthfeel of the yolk changes as a function of time at a constant temperature, see figure 31.

I believe that this study is a good practical approximation of texture design for a simple thermally treated egg yolk. It has a place in kitchens that practice science-based cooking as a smart tool that, ironically, stops the *cook-and-look* approach. The obvious challenge before us is to apply this analysis to the textural metamorphosis of the egg white. We can then combine the results and predict or design the texture of the perfect soft-cooked egg. If such a journey proves to be a platonic one, I would just scramble and cook the egg *inside* its shell and worry about only one thing: How will it taste?

Further Reading

Gadsby, Patricia. 2006. "Cooking for Eggheads." *Discover Magazine*, February. Available at http://discovermagazine.com/2006/feb/cooking-for-eggheads.

McGee, Harold. 2004. *On Food and Cooking: The Science and Lore of the Kitchen*, 70. New York: Scribner.

This, Hervé. 2006. *Molecular Gastronomy: Exploring the Science of Flavor*, 29–31. Translated by Malcolm DeBevoise. New York: Columbia University Press.

Vega, C., and R. Mercadé-Prieto. 2011. "Culinary Biophysics: On the Nature of the 6X°C Egg." *Food Biophysics* 6:152–159.

nineteen

KETCHUP AS TASTY SOFT MATTER

The Case of Xanthan Gum

THOMAS VILGIS

most people love tomato ketchup. It is tangy, sweet, and mouth pleasing. Our love for ketchup transcends flavor. It is loaded with nostalgia—the ultimate secret ingredient. I would venture to say that ketchup has gastronomic appeal. To make a case for this, let us look at ketchup in more detail. Specifically, let us analyze the particular sensations ketchup's ingredients evoke and explore their culinary possibilities. Simple as it looks, ketchup has a complex bouquet of aromas and an intricate mix of flavors, stimulating the salty, sour, sweet, and umami (or savory) taste receptors on first contact with the tongue. Above all, ketchup possesses that familiar and oh-so seductively smooth mouthfeel.

If you try the recipe for homemade ketchup, you will find the taste to be perfectly acceptable, but you will also find, well, that something is missing. Pour a small puddle of ketchup on a plate, and you will soon see a watery halo develop around the ketchup's outer edges (figure 32). This leaching out of water makes for a thinner, less creamy mouthfeel, drastically reducing the ketchup's richness of flavor. Tastants (that is, molecules that impart taste) leached into the water are more quickly released into the mouth. What causes this undesirable water loss and how can we resolve the problem?

Getting It Right by Physics

Imagine for a second that we were able to switch off our senses of taste and smell and perceive merely the texture of ketchup: we would feel only the sauce's thickness, or viscosity (a purely mechanical property touched on in chapter 18). Thickness has a strong impact on taste, as every chef intuitively knows. Thicker sauces remain in the mouth longer and release their flavors more slowly. When a sauce lingers in the mouth, taste receptors for sour, sweet, bitter, salty, and umami are under the influence of the taste molecules for a longer period. Moreover, the longer the ketchup stays in the mouth, the warmer it becomes and the more its volatile compounds escape, enhancing the overall flavor experience (for the role of thickness in the flavor perception of soups, see chapter 20).

HOMEMADE TOMATO KETCHUP

Let us first look at a tomato ketchup recipe. This is just one of thousands of easy home recipes you can make.

2¼ pounds (1 kg) tomatoes, diced
1 onion, diced
4 ounces (120 mL) white wine vinegar
1 ounce (30 g) sugar
10 g salt
1 red hot chili pepper, diced
10 grains of black pepper
5 grains allspice
¾ ounce (20 g) fresh thyme

Combine all the ingredients in a pot and simmer for 1 hour over low heat. Pass through a fine sieve, removing any large pieces, such as the tomato seeds and onions, to obtain a red liquid that is smooth and flavorful. Return the liquid to low heat and reduce it to the desired dense consistency. Fill clean bottles with the ketchup while the liquid is still very hot—this ensures an internal bottle temperature that inhibits the growth of bacteria. Seal the bottles; invert them and let them cool in that position. Enjoy your homemade ketchup!

Approximately 95 percent of ketchup is made of small molecules, such as water, vinegar, sugars, salts, and aromatic compounds. The other 5 percent consists of large molecules, such as polysaccharides, the majority of which are starch and fiber. These polysaccharides have the ability to hold a lot of water, and as a result, they thicken the ketchup. They owe this ability to their electrical charge, which binds water. For an illustration of how a polysaccharide becomes "dressed" with a shell of water, see figure 33.

Pectins, familiar to anyone who makes jam, are one example of such large molecules. Pectin is the "cement" that holds together the cell walls of most fruits and vegetables. It is released from the cell walls

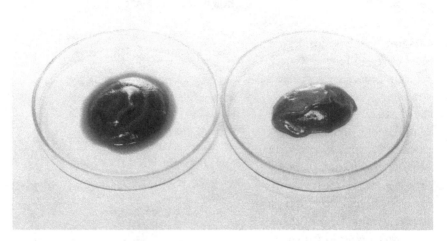

Figure 32 Homemade ketchup (*left*) gradually loses water, and eventually a liquid ring will form around the ketchup drop. A drop of ketchup made with xanthan gum (*right*) does not ooze water, better maintaining its shape. (Photograph courtesy of Thomas Jupa)

during cooking, but pectin amounts can vary depending on the fruit and its degree of ripeness. In the ketchup recipe already described, the pectin extracted from the tomatoes is obviously insufficient to hold on to all the water—hence the leaching. One solution would be to reduce the amount of water by further cooking. However, while this works, the texture suffers, becoming coarser and much less desirable. Another route is to add a few more large water-capturing molecules to the ketchup. I opted for the polysaccharide xanthan gum, so-called because it is produced from the fermentation of sugars by the bacterium *Xanthomonas campestris*. For the ketchup recipe I tested, I found that less than 0.5 gram of xanthan per 3½ ounces (100 g) of ketchup prevents most of the water loss. Not only that, xanthan yields a pleasantly smooth mouthfeel. What a remarkable molecule!

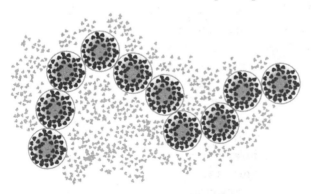

Figure 33 Thanks to its electrical charges, a polysaccharide chain can attract large amounts of water. These bound water molecules move more slowly than free water molecules in the surrounding liquid. Adding xanthan gum to ketchup therefore thickens it.

The Miracle of Xanthan Gum

Let us concentrate for a moment on the wonders of xanthan gum. What does this polysaccharide look like? At the molecular level, xanthan is a chain of up to 1 billion negatively charged sugarlike molecules. It forms a single helix, just like RNA. The molecule is similar to a rod: it is very long, straight, and difficult to bend.

One reason xanthan is useful for so many industrial and culinary applications is that it is not much affected by acidity, salt content, or even temperature, given its rodlike stiffness.

Zooming out a bit to observe the interactions between xanthan molecules in a liquid, we see that when two long charged rods come close together, they will strongly repel each other. If there are few rods in the solution, they can always avoid one another and move freely and independently. But if the number of rods in solution— their concentration—increases, movement becomes more and more difficult. At a high enough concentration, the rods tend to position themselves at right angles to one another as a result of the forces among them. This pretty much immobilizes them, also trapping the water between them. It is the formation of this network that marks the transition from a liquid to a gel-like system.

Now, imagine dabbing a french fry in ketchup made with xanthan gum and eating it. What happens to that blob of ketchup once it is inside your mouth? Chewing, of course, exerts force on the ketchup and the xanthan network in the gum. And because this mechanical force is large relative to the forces keeping the xanthan rods immobile, chewing remobilizes the rods, forcing them to slide along one another. This in turn frees up the movement of liquid and is perceived as thinning. The stronger the force upon the ketchup and its molecular xanthan rods, the thinner it will seem and the faster the rods will move.

To summarize: xanthan forms a network of immobilized rods, and this network is able to hold water tightly. If forced to flow, the rods change position easily; and the larger the force, the faster the ketchup will flow—that is, the thinner the ketchup will appear. Hence, the flow behavior of the ketchup, or any other food for that mater, is what constitutes its mouthfeel (as nicely demonstrated in chapters 18 and 20).

Xanthan's properties have some practical consequences in the kitchen. For instance, when thickening a liquid, it is often counterproductive to increase the concentration of xanthan much beyond the point at which the liquid no longer flows. In fact, overdosing promotes interactions between the rods and the rest of the ingredients, causing an unfortunate separation of the mixture into a rod-rich phase and water-rich phase. This creates unpleasantly slimy textures (a phenomenon analogous to that so elegantly explained in chapter 6).

Then there is the question of ketchup's pourability. When ketchup is at rest in its bottle—its contents undisturbed—it can be very difficult to pour, taking a bit of force to break the xanthan network and get the ketchup flowing. This *minimum* amount of force to create flow is what scientists refer to as yield stress. Once the yield stress is surpassed, the ketchup pours. As it flows, it thins, and as it thins, it moves faster. This is what vigorously shaking the bottle achieves. Hence a little rhyme, familiar to people of a certain age.

> Shake and shake
> the catsup bottle
> first none'll come
> and then a lot'll.

From Green Ketchup to Strawberry Ketchup and Beyond

With knowledge of how xanthan gum works, let us now prepare something completely different—a sour green ketchup. Take sour

Figure 34 A mixed-vegetable "tartlet" stabilized with xanthan gum.

gherkins (cornichons) and grind them together in a blender with spiced liquid, and strain to remove any large solid bits. Weigh enough xanthan gum so it represents about 1 percent of the final puree and rehydrate the powder with a portion of the puree's sour green liquid. Lumps might appear, but they are easily broken up with a hand mixer. Do not worry if air is incorporated because the bubbles will lend lightness and enhance the overall textural experience.

Any liquid that dissolves xanthan will do a similar job. Strawberry ketchup? No problem. Passion fruit? Go for it. Remember, xanthan is not especially sensitive to the salt, acidity, or sugar content of the liquid, so play away! Another recipe will get you started (figure 34).

MIXED-VEGETABLE TARTLET

With xanthan gum added, these "tartlets" can be made without the use of heat because the gum dissolves even in cold liquids.

7 ounces (200 g) ripe tomatoes
Salt and pepper
0.5 g xanthan gum
7 ounces (200 g) green beans, cut into small pieces (the lengths of which should equal their diameters)
3½ ounces (100 g) red and yellow bell pepper, skinned and diced into 4-mm cubes
1¾ ounces (50 g) fresh cilantro (coriander)

Using a small, sharp knife, make a small cross on the skin of each tomato. Place the tomatoes in boiling water for about 1 minute. Once cool, remove the skins and cut tomatoes into small dice. Reserve about 1½ ounces (50 mL) of the cooking water.

Season 2 ounces (60 mL) of the cooking water with salt and pepper and mix in the xanthan. Boil the green bean pieces for 7 minutes and drain them in a sieve.

Mix all the diced vegetables and cilantro together, and add the "xanthanized" cooking water.

Use individual 3-inch (8 cm) metal rings to quickly form the "tartlets" on individual plates. Take off the rings and dress with a vinaigrette of your choice.

Further Reading

Sanderson, G. R. 1981. "Application of Xanthan Gum." *British Polymer Journal* 13:71–75.

Vilgis, Thomas. 2005. *Molekularküche: Das Kochbuch*. Wiesbaden: Tre Torri.

——. 2010. *Das Molekül-Menü: Molekulares Wissen für kreative Köche*. Stuttgart: Hirzel.

twenty

TASTE AND MOUTHFEEL OF SOUPS AND SAUCES

JOHN R. MITCHELL

in traditional cuisine, soups and sauces are thickened with starch-based ingredients, such as wheat flour. In today's highly evolved culinary world, the use of hydrocolloids for texture development is more the rule than the exception (chapter 4). Although it is possible to obtain a similar degree of thickening[1] in soups and sauces with nonstarch polysaccharides—such as guar gum, xanthan gum, locust bean gum (LBG), and carboxymethylcellulose—the mouthfeel and flavor are often not as good as traditional starch-based ingredients. I believe the key to these differences can be found in the way the food mixes with saliva in the mouth. Oddly enough, gelatin, which is neither a starch nor a polysaccharide but a protein, has maintained its preeminent position as a thickener of stocks and sauces and as a gelling agent in most terrines and desserts.

The fact that gelatin is still the ingredient of choice might be explained by how it behaves in the mouth. Concentrated, highly viscous gelatin solutions mix very quickly with water and saliva at body temperature. By this, I mean that the thick sensation readily disappears as

1 Most liquid foods are non-Newtonian; that is, their viscosity decreases the faster they are stirred. To design foods with similar degrees of thickening using the measured viscosity as the reference parameter, one needs the "shear rate" (the rate at which layers of fluid slide over one another) within the mouth. This is often taken as 50 s^{-1}, but this value will depend on the food.

the saliva dilutes the concentrated gelatin solution. Something similar happens with sauces thickened with starch, like béchamel. In contrast, thick, viscous solutions of nonstarch polysaccharides often mix very poorly.

This poor mixing has two undesirable consequences:

1. Taste molecules, particularly salt and sugars, will remain in the poorly mixed solution and be swallowed before their concentration equilibrates with saliva in the mouth, resulting in inhibited saltiness or sweetness.
2. If a viscous solution mixes poorly with saliva, it will dilute very slowly, giving an undesirable mouthfeel, which can be described, depending on the polysaccharide, as slimy, mouth coating, clingy, and the like. What is desirable is a less viscous solution by dilution with saliva that makes swallowing as easy as possible.

In brief, thickening achieved with nonstarch polysaccharides seems to be detrimental to taste and mouthfeel when compared with traditional starch-based approaches. Here is why.

C^* Concentration

As described in chapter 19, a polysaccharide molecule, like xanthan gum, can be visualized as an expanded coil or rod that has plenty of room to bind to large amounts of water. In so binding, the solutions that contain them thicken. As the concentration of the polysaccharide increases, so does the viscosity of the solution. However, given the large size of the coils, a concentration is soon reached at which the coils start touching one another. At this point, the viscosity of the solution dramatically increases. This is what scientists call the c^* (c star) concentration. Interestingly, for solutions of nonstarch polysaccharides, poor mixing behavior seems to occur at concentrations above c^*. Because the polysaccharide coils bind so much water, the c^* concentration is low—for example, that of guar gum is about 0.5 percent. The exact value depends on the molecular weight of the polysaccharide; therefore, xanthan, guar, and locust bean gums have different c^* concentrations because their molecular weights differ. As more polysaccharide is added past the c^* concentration, the molecules go from just touching one another to overlapping and entangling. When water, or saliva, is added to a solution that is above the c* concentration, dilution takes considerably longer to occur because water has a hard time flowing into the network of entangled molecules.

The decrease in taste perception above c* and the probable role of dilution and mixing was first recognized by Zoë V. Baines and Edwin R. Morris (1987). Diffusion will be the final transport mechanism by which tastants reach the taste receptors in the mouth. However, diffusion processes are slow, so dilution and mixing will play key roles in the transport of the tastants to the taste receptors.

Differences in Mixing Behavior

The efficiency of mixing can be illustrated by simple experiments in any kitchen. The key is to add a small quantity of color to the viscous solution, which is then mixed with more water and stirred by hand for a few seconds.

Figure 35 compares the appearance of a wheat starch–thickened solution with a solution thickened using hydroxypropylmethylcellulose, a nonstarch polysaccharide, following mixing with water. The initial shear viscosities of the two solutions were the same. The superior mixing of the wheat-starch system is obvious. The LBG solutions were made at concentrations below and above their c* concentra-

Figure 35 Appearance after mixing of colored wheat starch (*left*) and hydroxypropylmethylcellulose (*right*) "solutions" in water. The initial viscosity of the two thickened solutions is the same (380 mPa.s at a shear rate of 50 s⁻¹). (Reproduced with permission from A.-L. Ferry, J. Hort, J. R. Mitchell, D. J. Cook, S. Lagarrigue, and B. Vallès Pàmies, "Viscosity and Flavour Perception: Why Is Starch Different from Hydrocolloids?" *Food Hydrocolloids* 20 [2006]: 855–862)

tions. It is noteworthy that in contrast to LBG, molten gelatin at high concentrations and high viscosities retains the ability to mix well. It is likely that this very efficient mixing when combined with gelatin's well-known melt-in-the-mouth property is why a gelatin gel has such good mouthfeel.

The experiment demonstrates the relevance of c^* in the mixing behavior of the viscous LBG solutions. However, scientists still do not fully understand why gelatin solutions mix so well. The polymer coils appear not to entangle in the same way as linear polysaccharides, which explains why highly concentrated solutions show Newtonian behavior; that is, their viscosity remains constant regardless of how fast they are stirred.

Treat Starch Gently

The microstructure of starch-thickened solutions is generally a suspension of swollen granules (see figure 11), and these granules do not show the same entanglement interaction that polymers in solution do. If this granular structure is disrupted by excess shearing, for example, the starch will begin to show poor mixing, resulting in the reduced taste release and long, stringy textures characteristic of nonstarch polysaccharides solutions. Waxy cornstarch consists almost entirely of the branched polysaccharide amylopectin, and it is disrupted very easily upon heating and shearing. To prevent this, the industry physically modifies or chemically cross-links the granules. Figure 36 shows the difference in the microstructure and mixing behavior between a native cornstarch and a physically modified waxy cornstarch cooked under the same conditions, heating to 203°F [95°C]. They are moderately stirred and the temperature is maintained for 2½ minutes, followed by cooling.

As the mixing behavior of these starch solutions is relevant to taste release in the mouth, I was curious about the effect of amylase, a starch-degrading enzyme found in saliva, on the mixing behavior of native and modified waxy cornstarch. Within a few seconds, amylase acts fast enough to lower the viscosity of starch suspensions; however, because there is a partial conversion from a granular structure to one consisting of starch polymers in solution, the mixing behavior and taste perception actually worsen despite the decrease in viscosity.

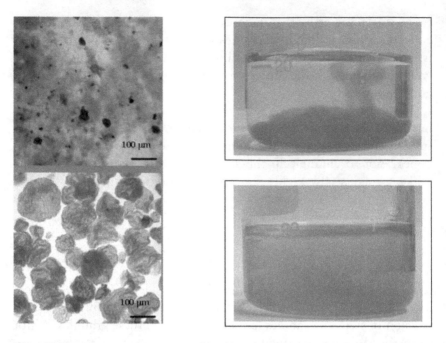

Figure 36 Microstructure and appearance after mixing of native (*top*) and physically modified waxy cornstarch (*bottom*). Micrographs of starch granules obtained after staining with iodine. (Reproduced with permission from Ferry et al., "Viscosity and Flavour Perception"; and A.-L. Ferry, J. R. Mitchell, J. Hort, S. E. Hill, A. J. Taylor, S. Lagarrigue, and B. Vallès Pàmies, "In-Mouth Amylase Activity Can Reduce Perception of Saltiness in Starch-Thickened Foods," *Journal of Agricultural and Food Chemistry* 54 [2006]: 8869–8873)

The Soup Party

There is sensory evidence that people with higher levels of in-mouth amylase activity perceive saltiness to a lesser degree. It would stand to reason, then, that if an unsalted wheat flour–thickened soup were served at a dinner party, and the guests were given the opportunity to add salt to taste, those with high levels of in-mouth amylase activity would add more salt. The observation that amylase activity lowers taste perception is at first sight counterintuitive. I initially expected that the reduction in starch viscosity as a result of amylase activity would improve taste perception. This is because it has been accepted for a long time that taste perception decreases with increasing viscosity. The hypothesis based on the data presented here is that amylase partially converts the sauce microstructure from a suspension of swollen particles that mixes well to a polymer solution that mixes poorly. It is this reduction in mixing ability that overrides the benefits, to taste perception, of the viscosity reduction after amylase addition.

The Rheological Challenge

Plenty of studies exist that look at the relationship between viscosity and the perception of foods. The simple experiments I describe suggest that at high viscosities, sauces and soups thickened with nonstarch polysaccharides or starch that is overcooked mix very differently with saliva compared with sauces and soups containing starches in which the swollen granules remain intact. This provides a probable mechanism for the difference in perception of "short" (good-mixing) and "long" (poor-mixing) textures. The view was often taken that taste perception decreases with increasing viscosity. As previously discussed, there are situations in which taste perception decreases with decreasing viscosity. What is required is a parameter or combination of rheological parameters that can be used to predict taste perception. Although one of these parameters will be the shear viscosity, other factors will probably be required. The extensional viscosity (how a liquid responds to stretching), although not easy to measure, may supply the missing dimension. Predicting perception of foods from rheological measurements remains a challenge. We not only need to go beyond the measurement of shear viscosity but we must also consider the changes that take place as a result of mixing the food with saliva.

Acknowledgments

The results shown in this contribution have been obtained by Ann-Laure Ferry, Andreas Koliandris, and Angela Lee, all of whom have made the development of these ideas possible. I also gratefully acknowledge the input of my academic colleagues Joanne Hort, Sandra Hill, Bettina Wolf, Dave Cook, and Andy Taylor of the University of Nottingham and my wife, Margaret Hill, who through her doctoral work started my interest in understanding the excellent oral perception that starch-thickened products display.

Further Reading

Baines, Z. V., and E. R. Morris. 1987. "Flavour/Taste Perception in Thickened Systems: The Effect of Guar Above and Below c*." *Food Hydrocolloids* 1:197–205.

Ferry, A.-L., J. Hort, J. R. Mitchell, D. J. Cook, S. Lagarrigue, and B. Vallès Pà-mies. 2006. "Viscosity and Flavour Perception: Why Is Starch Different from Hydrocolloids?" *Food Hydrocolloids* 20:855–862.

Ferry, A.-L., J. Hort, J. R. Mitchell, S. Lagarrigue, and B. Vallès Pàmies. 2004. "Effect of Amylase Activity on Starch Paste Viscosity and Its Implications for Flavour Perception." *Journal of Texture Studies* 35:511–524.

Ferry, A.-L., J. R. Mitchell, J. Hort, S. E. Hill, A. J. Taylor, S. Lagarrigue, and B. Vallès Pàmies. 2006. "In-Mouth Amylase Activity Can Reduce Perception of Saltiness in Starch-Thickened Foods." *Journal of Agricultural and Food Chemistry* 54:8869–8873.

Koliandris, A., A. Lee, A.-L. Ferry, S. Hill, and J. Mitchell. 2008. "Relationship Between Structure of Hydrocolloid Gels and Solutions and Flavour Release." *Food Hydrocolloids* 22:623–630.

Mitchell, John R., and Bettina Wolf. 2011. "Relationship Between Food Rheology and Perception." In *Practical Food Rheology: An Interpretive Approach*, edited by Ian T. Norton, Fotios Spyropoulus, and Philip Cox, 173–192. London: Wiley-Blackwell.

twenty-one

PLAYING WITH SOUND

Crispy Crusts

PAULA VARELA AND SUSANA FISZMAN

according to the quiche recipe by Dorie Greens-
pan (2009), "as soon as the crust comes out of the oven, lightly beat
an egg white with a fork and brush the white over the inside of the
crust. . . . [I]t will provide a kind of waterproof lining between the
crust and the quiche filling. Quiches are so much nicer when you can
pair their soft, creamy custard with the slight crunch of a crust."

The fact is that baked dough products, like quiche and pizza,
deteriorate rapidly after they are removed from the oven. Absorption of water into the crust is the main problem, which contributes
to the deterioration of the crispy texture and the presence of an
unpleasantly soggy crust. Let us look at how crispness is achieved,
perceived, and analyzed in foods with a crispy crust and a soft,
moist interior.

Crispy or Crunchy Texture?

From a scientific point of view, texture is related to the structural,
mechanical, and surface properties of food. Texture is detected by
seeing, hearing, touching, and kinesthetics (the ability of the body to
sense its position, location, orientation, and movement—and those of

its parts). Only human beings can perceive and describe food texture, and we detect it as one integrated sensation, which nevertheless comes from the various senses—all stimulated by the eating process. The perception of food is gleaned from the information collected by the senses through the process of eating. This includes observing the food product outside or inside its container and handling the food item before consumption (cutting roast beef or pouring soup), as well as actually eating it. Texture plays an important role in determining likes and dislikes, and its importance is accentuated when expectations are violated, for example, when one tastes meat that is not tender, lumpy puree, or ice cream that is crystallized rather than creamy.

Crispness and crunchiness are not so much characteristics of the food item as they are aspects of its structure and the physical state of its components. Whether these two English terms refer to different concepts is still open to question (for a discussion of the sensory experience of food and crispy–crunchy–crackly foods in particular, see chapter 2). In other languages (for example, Spanish and French), similar questions arise. What crispness and crunchiness actually mean is not clear because the perception of crispness and crunchiness is the sum of the information received by several senses (such as sight, hearing, and touch). Crispy is, for instance, defined by Laurence Fillion and David Kilcast (2002) as "a light and thin texture producing a sharp clean break with a high-pitched sound when a force is applied, mainly during the first bite with the front teeth." Crunchy is described as "a hard and dense texture that fractures without prior deformation." Lisa M. Duizer, Osvaldo H. Campanella, and Geoff R. G. Barnes (1998) define crispness as "a combination of the noise produced and the breakdown of the product as it is bitten entirely through with the back molars." Confusion abounds! Zata Vickers (1984) initially suggested that the pitch of the sound was the distinguishing factor between crispness and crunchiness; however, the same author later said that "most foods were about as crispy as crunchy." Therefore, for our purposes, we will speak exclusively in terms of crispness, while not forgetting that crispness and crunchiness are complex concepts with interpretations that depend on the food item, the background of the eater, and their food culture.

Foods that are crispy because they have a crust tend to be fresh, recently baked products (like French bread). Crispness can take the form of an added element, such as nuts in salads or croutons in soups. These added "touches" create pleasing, creative, or surprising textural combinations. Recent developments in the art of cooking impart new textures to traditional preparations or combine various textures in the same dish made with the same raw materials. For example, we have seen these items used in combination with various other foods: orange gelatin on sliced fresh oranges, caramelized orange peel in orange sherbet, and candied orange peel in orange ice cream.

In a study performed in our lab, a total of two hundred consumers were interviewed about crispy food (Varela et al. 2007). Foods these consumers most frequently described as crispy were so-called dry–crisp products, such as toast, cookies (biscuits), potato chips (crisps), and breakfast cereals; these foods are firm and brittle. The second most mentioned group of crispy foods was what could be called wet–crisp products, which include raw or hardly cooked vegetables and raw fruits, such as apples. In wet–crisp foods, the crispy characteristic comes from the water pressure inside the plant cells, which are what make up the vegetable or fruit. A third group was foods with a crust, which, as the name already implies, are foods with a dry, hardened exterior and a soft, high-water-content core. Familiar examples are french fries (chips), French-style bread, and battered-and-breaded deep-fried foods. Also included in this group were foods that combine a crispy layer and a high-water-content topping. The best-known examples are pizzas, pies, quiches, and tarts. For reasons we will discuss shortly, retaining the crispy character of the dry part is very difficult.

How Water Is Related to Crispness

All products with a crust present common characteristics and therefore challenges because of their mixed structure, normally having a high-water-content, soft, and deformable interior, surrounded by, or attached to, a dry, firm, and brittle crust. The key issue in these kinds of products is how to maintain the crispy character after preparation; in general, the loss of crispness is due to the diffusion of water from the high-water-content part to the low-water-content, crispy part.

Water content in any given sample indicates how "humid" the food is. However, it is more accurate to speak about water activity (a_w). The water activity is a simple thermodynamic measure of the dryness of food. Water activity is related to how free, available, or bound the water is. In addition to the water content, water activity values depend on the concentration and type of dissolved substances in the food—mainly sugars and salt, the existence and distribution of pores, and the relation of water to the "thirsty" sites of the food matrix. Proteins, carbohydrates, and starches possess many sites to which water molecules can strongly bind. Critical water activity values are those in which the products become sensorially unacceptable because they completely lose their crispness.

Water diffusion from moist (high a_w) to dry (low a_w) layers causes moistening of the components of the crispy parts. At a critical water activity value, further movement of water causes a change from a glassy state (in which the materials behave as hard and brittle) to a rubbery state (in which the materials behave as leathery, soft, or sticky). As the food is transitioning between the glassy and rubbery states, a loss of crispness takes place.

Battered and Breaded Products

Tempura or battered-and-breaded fried foods—like fish, seafood, poultry, cheese, and vegetables—are good examples of foods with crisp external crusts. They are favored and appreciated by consumers. Coated fried foods are dipped in a flour-based batter before they are fried. Because battering and frying have been traditional methods for preparing foods, empiricism has dominated their application for centuries. Although considerable geographical variations occasionally exist because of the raw materials available, versions of batter-coated or breaded foods are found in the traditional or regional cuisines of practically every part of the world.

Tempura-Type Coatings

In one type of coating, known as tempura, the liquid batter comes into direct contact with the hot oil, which makes it coagulate around the piece of food, forming a crispy crust. All the final quality characteristics of the tempura-type coated food largely depend on a good formulation of the ingredients that constitute the raw batter. The list of ingredients used has become much longer than just wheat flour and water: different starches, gums, milk, seasonings, and many other items are added. The behavior of each ingredient is very different, determining the final performance of the product.

This batter is a single outer covering for the substrate. Unlike adhesion batters (as we will see), this type of batter normally contains leavening agents (beer, for example) that contribute to the formation of small bubbles (carbon dioxide). Therefore, the batter expands when fried, developing a number of larger gas cells and, consequently, a spongy structure. Celebrated chef Heston Blumenthal of the Fat Duck in Berkshire, England, created a novel way to get significantly crisper fried fish. This was done by manipulating the type of starch used—by nucleating the batter in a syphon with nitrous oxide and by incorporating vodka (alcohol takes far less energy to evaporate than does water). All this rendered an irregular and rather unique crust while ensuring that the fish did not overcook (for a nonsyphoned, but nonetheless crisp, battered fish recipe, see chapter 2).

An emulsifier, such as the lecithin present in egg yolk, can keep the growing bubbles from collapsing or bursting. During frying, the batter foam dries out and takes on a completely solid and sponge-like structure. The use of a leavening agent reduces the density (less weight for the same volume) and increases the volume of the coating so that it is lighter on the tongue.

The leavening agent produces gas. This helps aerate the structure; however, only when small bubbles are already present in the batter will the released gas be captured by them—normally, no new bubbles are formed. Instead, those bubbles that are already present grow. This is why the initial beating stage is very important in this type of batter. The aeration caused by the leavening agent contributes to crispness and facilitates water loss (in the form of steam) during frying. Steam also helps to expand the coating.

The structural characteristics of tempura-like batter must be such that they lend to the fried product a uniform external layer with good adhesion to the substrate—and good coverage. Tempura-type batters form a crispy, continuous, aerated, and uniform layer over the food substrate that constitutes the batter's final aspect. They protect the natural juices of foods, thereby ensuring a final product that is tender and juicy on the inside and crispy on the outside.

Adhesion Batters

Another type of batter is one that acts as the glue to an external layer of bread crumbs, creating a battered-and-breaded final product. The choice of the batter ingredients is not as delicate a matter as in tempura-type products. Essentially, the batter needs to act as a good adhesive—and, ideally, it should not be distinguishable during consumption. The crispy characteristics of the end product in this case mainly depend on the bread-crumb coating: the bread or grains from which it is made, the shape and size of the crumbs, and the crumbs' regularity, degree of toasting, and so on. After a few minutes of frying, coated foods have a pleasant golden-colored exterior with a crunchy texture, whereas the interior usually remains tender and juicy. These are the characteristics that make these products so appetizing.

Cooking Battered Food Items

Frying is the most common method for cooking or reheating tempura or battered-and-breaded foods. Apart from a variety of chemical reactions occurring, several changes take place in the frying process, such as gelatinization of starch, denaturation of protein, and reduction of moisture (the product is dehydrated until it provides a crispy texture).

One problem associated with the consumption of battered-and-breaded deep-fried foods is the great amount of oil absorbed during the frying process. Recently, there has been a trend to reduce the fat content in fried foods by changing the formulations or developing new cooking methods to avoid one of the frying steps. A good solution for this is the use of prefried frozen or refrigerated battered or battered-and-breaded food pieces. For instance, only 30 seconds of prefrying is required in the case of battered shrimp. Such products may then be finished by alternative cooking methods; this constitutes a useful option for caterers or hospitality kitchens, for example, with the added advantage of leading to a lower fat content than the equivalent fully deep-fried products.

The oven is increasingly used as an alternative cooking or reheating method for prefried battered and breaded foods. Baking is a good way to avoid the excessive absorption of fat that occurs in deep frying. New methods in microwave heating are another option. The preparation of battered-and-breaded food items in standard microwave cooking notoriously leads to serious texture problems: sogginess, lack of crispness, and lack of browning. This is because the microwave radiation heats up the moist interior of the food item. It thus drives off the water (in the form of vapor), forcing it to move from the interior and outward to the crust, where the water condenses because the crust is still relatively cold.

However, new technologies offer a way to specifically heat the *surfaces* of food items in microwave cooking. One of the more recent methods that can be applied to cook or reheat prefried products in the microwave uses susceptor materials. The name is derived from *susceptance*, a property of certain materials that gives them the ability to convert electromagnetic energy (from the microwave) into heat. In fact, the popcorn susceptor bag, in which the microwave temperature ultimately reached is high enough to cause corn kernels to pop, is a well-known example. Susceptor materials are metallized plastic films laminated with paperboard; the heat generated in the susceptor material during microwaving is transferred to the product, creating areas hot enough to evaporate water and render the food surface crisp. Susceptors can work reasonably well as long as the surface of the food item is in contact with or very close to the susceptor material (figure 37).

Figure 37 Nugget surrounded by the susceptor casing. The external surface is made of cardboard; the internal surface is constructed of a specially designed material that acts by heating up the sample surface through radiant heating.

Analyzing Crispness

In products with a crust, what we perceive as texture in the outer layer during eating, be it soggy or crispy, is certainly a result of the characteristics of the crust. The perception of crispness is partly related to auditory sensations, as all crispy foods are noisy when eaten. Scientifically, the study of the crispy or crunchy textures can be performed through recording the sounds emitted by the food piece while it is compressed or cut (for a detailed explanation of how this is done, see chapter 2).

Case Study: Prefried Chicken Nuggets Cooked by Classical and New Methods

The same prefried chicken nuggets were cooked by four procedures:

- Deep-frying (360°F [180°C]) in a domestic fryer (3 quarts [3 L] oil; 3 minutes)
- Electric oven (440°F [225°C]) with convection (11 minutes)
- Microwave oven (700 watts; 1 minute, 15 seconds)
- Microwave oven (700 watts; 1 minute, 15 seconds), with susceptor material (see figure 37)

Crispness is strongly linked to auditory sensations. Sounds contain important information related to the fracture properties of crispy foods. Not surprisingly, the instrumental methods developed by scientists to evaluate crispy foods in the laboratory have focused on measuring the sound they emit during fracture.

In this example, a microphone was used to record the sounds emitted while the sample was cut with a plastic blade.

The results can be evaluated by observing the curves, which relate the force required to cut the products with the sound they emit during cutting.

- *Deep-fried samples.* When the samples are very crispy, as indeed they were in this case, the curves obtained are highly jagged, with lots of peaks (figure 38a). This means that small cracks or fractures occur in the nugget crust as the blade penetrates the samples (imitating the teeth-biting action) and sudden drops of force take place producing audible noises.

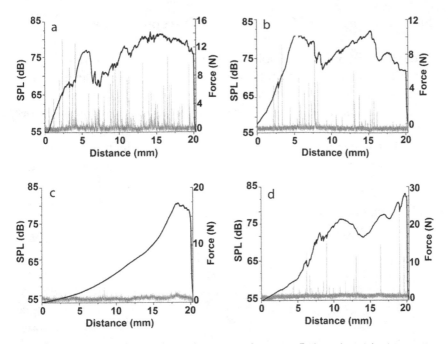

Figure 38 Force and noise emitted during the cutting of a nugget. Each graph contains two curves: the black one represents the force, and the gray one represents the noise. The force is measured in newtons (N) and the noise level is expressed as sound pressure level (SPL) in decibels (dB): (a) deep-fried sample; (b) oven sample; (c) microwave sample; (d) microwave plus susceptor sample.

• *Oven-cooked samples.* These samples show a curve with fewer force peaks. The force values are higher because a harder, drier product is obtained. This is caused by the greater water loss during the long exposure to the high oven temperature (figure 38*b*).

• *Microwaved samples.* The force plot obtained from microwave cooking is dramatically different from the others (figure 38*c*): it does not present peaks in force, meaning that no fracture events have happened. These samples obviously were not at all crispy but gummy and tough. The toughness is reflected in the force values, which are higher than in the deep-fried samples, indicating tougher pieces of food.

• *Susceptor-microwaved samples.* A noticeable improvement in crispness is seen when the susceptor material surrounds the product: some ups and downs appear in the force, together with some sound events (figure 38*d*). There was an enhancement of the crispy character, but these nuggets were still not as good as the deep-fried samples.

How a Microwave Heats and Why Susceptors Work

In deep-frying or in oven cooking, the water evaporation occurs mainly at the surface because heat comes from outside (from the hot oil or from the hot oven walls and air). Nuggets (or any other food item) heated in a microwave oven undergo internal water evaporation because the microwave energy heats all the water at the same time; this inner hot water tends to escape to the surface, which is surrounded by *cold* air (in a microwave oven, the air does not heat), causing a moistening of the crust.

When a nugget is microwaved in a susceptor material, the heat that the material generates (with the microwaves' help) allows the water to evaporate and the nugget surface to dry. The susceptor material therefore acts as a small oven. The use of susceptors constitutes the first attempt at preventing sogginess in microwaved food. There are already a number of industrial food products for home cooking that make use of this development, focusing on ready-to-heat crusted or composite foods for the microwave (sandwiches, pastries, and pizzas, for example).

The use of susceptors is just one example of the possibilities that food science and technology offer in the vast panorama of improving crispness in battered-and-breaded food. Much research into microwaving and other

---

CRISPY BATTERED CHICKEN THIGHS

4 chicken thighs
3½ ounces (100 g) plus 10 g all-purpose (wheat) flour
Salt, baking powder, and black pepper
2 egg whites
4 ounces (125 mL) cold sparkling water*

Preheat the oil to 390°F (200°C). Pat the chicken thighs dry with a paper towel. Put the 10 g of extra flour on a plate and use it to coat each thigh, patting off the excess. Mix together the remainder of the flour, a pinch of salt, a pinch of baking powder, and some pepper. Lightly whisk the egg whites until bubbly but not stiff. Pour the sparkling water into the flour mix, whisking gently and briefly. Gently stir in the whisked egg whites just to mix. Retain as many bubbles as possible so that the batter stays light.

Dip two thighs in the batter to coat, let the excess drip off, and then immerse them in the hot oil (390°F [200°C]) using a slotted spatula. Fry for 5 to 6 minutes, making sure the oil temperature remains constant. When the batter is set, turn the pieces of chicken over and cook until they are an even golden brown. Lift out with the spatula and drain on a paper towel. Check that the oil has returned to 390°F (200°C), and then repeat with the remaining chicken thighs.

* The bubbles from the sparkling water provide many tiny bubbles that act as nuclei. These grow in size when the carbon dioxide generated from the baking powder is released. Together, the bubbles and carbon dioxide contribute to the structure of the batter during frying, producing a spongelike, crispy texture.

cooking methods is yet to be done, with many questions still to be answered, such as how effective could ovens that combine infrared heating and microwaves be?

Other ways to enhance the crispness of coated foods involve the use of ingredients such as coarser bread crumbs, mixes of different crumb sizes, dried Japanese bread crumbs, whole grains, multigrains, seeds, or nuts incorporated into the batter. Also, starches lend crispness to batter coatings, particularly those coming from new vegetal sources, like sago, new rice cultivars, and the like. But the main question is, Are cooks ready to follow each of these crispy trails?

Acknowledgments

The authors are grateful to the Ministerio de Ciencia e Innovación of the Spanish government for its financial support (AGL 2009-12785-C02-01) and to INDAGA network (AGL2009-05765-E).

Further Reading

Duzier, L. M., O. H. Campanella, and G. R. G. Barnes. 1998. "Sensory, Instrumental, and Acoustic Characteristics of Extruded Snack Food Products." *Journal of Texture Studies* 29:397–411.

Fillion, L., and D. Kilcast. 2002. "Consumer Perception of Crispness and Crunchiness in Fruits and Vegetables." *Food Quality and Preference* 13:23–29.

Greenspan, D. 2009. "Quiche: And Now for the Crust." Available at http://www.doriegreenspan.com/print/2009/02/quiche-and-now-for-the-crust.html.

Varela, P., A. Salvador, A. Gámbaro, and S. Fiszman. 2007. "Texture Concepts for Consumers: A Better Understanding of Crispy–Crunchy Sensory Perception." *European Food Research and Technology* 226:1081–1090.

Vickers, Z. 1984. "Crispness and Crunchiness: A Difference in Pitch?" *Journal of Texture Studies* 15:157–163.

twenty-two

BAKED ALASKA AND FROZEN FLORIDA

On the Physics of Heat Transfer

ADAM BURBIDGE

have you ever watched chefs in a Chinese restaurant prepare a meal? Impressive, is it not? A couple of minutes of frantic stir-frying in a wok on a fierce gas flame and—presto—a delicious meal! Wish you could achieve the same effect at home? Dinner in three minutes would be a cool trick and really impress the kids! To accomplish this and similar feats, and gain some knowledge of kitchen science in the process, it is necessary to understand a little bit about the physics of heat transfer, which I provide here in the form of a brief introduction to the concept. Then I describe and explain what this means in the context of real cooking. With a better grasp of the underlying physics, you will be better equipped to try some of the recipes I provide, which offer a few interesting twists on some delicious classics.

An Easy-to-Swallow Introduction to Heat Transfer

Heat transfer refers to the motion of heat, which is a form of energy that moves through a material, which in this case is food. We do not directly perceive energy but instead sense how much of it is present by means of temperature; hot things have a lot of energy relative to cold things. Because heat is just a form of energy, we can raise the temperature of something by adding energy. There are, of course, numerous ways to do this, the most obvious being exposure to something hotter (such as air in a heated oven), but other less obvious methods are also effective (such as friction—Boy Scouts will enthusiastically explain to you that you can start a fire by rubbing two sticks together).

There are four basic methods by which heat travels: conduction, convection, radiation, and volume heating. Conduction is the transfer of heat from a hot object to a cold object, or from a hot region to a cold region within the same object. For example, if I heat a brick at one end, the whole brick will eventually get hot.

With conduction, there is no movement between the objects—the bricks are stationary. But with convection, the movement of heat comes from the movement of the material. For example, I pump hot water down a pipe, and heat is carried along the pipe in the moving water. We can see from this that conduction requires a temperature difference, but convection does not. Radiation[1] is important at high temperatures and travels in waves in a similar fashion to light—this is how the sun heats the earth and explains why you do not get sunburned on the side that is not exposed to the sun. It also explains why you need to turn a roasting chicken over on a barbecue because the hot coals transfer a large amount of their heat through radiation. Radiation does not require contact and can travel through a vacuum, as it does in outer space. The final way heat is transferred is by volume heating. This is essentially how microwave ovens cook food: microwaves heat the whole volume of the material simultaneously.[2]

Each material—in this case, food material—possesses a number of intrinsic properties that allow us to characterize its behavior when compared with other materials. Such properties include:

- *Specific heat capacity* (C_p). How much heat you need to supply to raise the temperature of a fixed mass of something (usually 1 kg)

1 This has nothing to do with fallout, nuclear power, or the like. It is just a technical term to describe how heat moves as thermal waves.

2 More specifically, microwave ovens use a process called dielectric heating. Simply speaking, the magnetron emits microwaves, which oscillate at very high frequencies and interact with any polar molecules, such as water, in the material being cooked. The fast motion of the polar molecules dissipates a lot of molecular-scale frictional energy, which heats the material at a rate proportional to the local concentration of water. Of course, areas with more or less water will be heated at a greater or lesser rate than the average, depending on the local water content.

	ρ (kg/m³)	C_p (J/kgK)	k (W/mK)	α (mm²/s)
Air	1.2	1,005	0.026	21.6
Apples	850	3,900	0.39	0.12
Beef	1,100	3,400	0.48	0.13
Bone	440	830	0.55	1.5
Bread	300	2,900	0.05	0.06
Butter	911	2,100	0.20	0.11
Cod	1,000	3,600	0.54	0.15
Ice cream[a]	550 (1,100)	3,100	1.1 (0.4)	0.24 (0.32)
Meringue[b]	100	400	0.05	1.3
Nuts	700	1,200	0.2	0.24
Peas	820	3,400	0.31	0.11
Water	1,000	4,186	0.58	0.14

Note: ρ, density; C_p, specific heat capacity; k, thermal conductivity; α, thermal diffusivity
[a]The data for ice cream is for the frozen state; the figure in parentheses refers to the degassed mix at room temperature.
[b]The value for meringue is estimated from a sample at 1,000 percent overrun.

by 1 degree Celsius. The larger the heat capacity of something, the more difficult it is to heat up.

• *Thermal conductivity (k)*. How good a conductor something is. High values mean the material is a good conductor, low values a poor one (that is, the material is an insulator). It is important to note that this describes only conductivity and provides no information about the other heat-transfer methods.

• *Density (ρ)*. The weight of a fixed volume of something. For example, a liter of air weighs less than a liter of water; hence, air is less dense than water.

• *Thermal diffusivity (α)*. The ratio of how well something conducts heat to how easy it is to heat up. High α materials tend to quickly and easily reach a constant temperature; low α materials tend to maintain large temperature differences for long periods of time.

Some typical values for these properties are given in table 1.

A few more scientific concepts are worth mentioning before we step into the kitchen.

• *Boiling and phase change*. Boiling occurs when a liquid material is sufficiently heated to spontaneously form vapor bubbles. The change from liquid to gas is called a phase transition. The heat required for this transition to occur, called latent heat, is very large compared with the heat capacity of the material being boiled, so it follows that the temperature of the material does not increase during the phase transition. A similar thing occurs when a solid material is heated to its melting point: the temperature remains almost constant until all of the material has melted.

• *Cooking as a chemical reaction.* *Cooking* is a general term that relates not just to physical changes but to the progression of a huge number of chemical reactions in food, including those involved in browning and flavor development (the Maillard reaction). Reaction kinetics is obviously a complex topic, but in general, the rate of the reactions that occur during cooking increases rapidly with rising temperature (for a vivid example of this phenomenon, see chapter 18).

• *Relaxation time.* This naturally leads us to the concept of the so-called relaxation time—the time it takes for a food to reach a constant and uniform temperature throughout. A sizable leg of lamb cooking in the oven could have a relaxation time as long as an hour. In themselves, relaxation times are not especially interesting, but they provide a ready understanding of processing time in cooking and serve to delineate fast from slow cooking. Fast cooking occurs when the total cooking process is shorter than the relaxation time, as in making toast on a grill (that is, just toasting the outside of the bread); slow cooking occurs when the cooking time is much longer than the relaxation time, as in roasting a turkey.

• *Physical properties of ingredients.* The diffusivity value (α) for various materials, ranging from solid (beef) to liquid (water) to gas (air) are shown in table 1. Air (a high α material) reacts very fast, which is why your oven works. A pan of water reacts about one hundred times more slowly, which is one reason, although perhaps not the most obvious one, your oven is not full of water. Of course, there are other considerations related to convective heat transfer that arise because air (gas) or water (liquid) can move. Generally, for large bodies of gas or liquid, convective effects are dominant. These allow us to fly gliders on "thermals," which are convectively driven winds. Solid materials cannot move or flow by themselves as water does, so there are no convective effects. As a consequence, solids tend to be less efficient at transferring heat over large distances than are liquids.

Bread (low α) has a very small thermal diffusivity, meaning it is slow to heat through evenly—so it maintains temperature differences for long periods of time. This is why toast can be crisp on the outside while remaining moist on the inside, and it is why, at least in theory, you can gain escape from a burning building by covering yourself in sliced bread.[3] Table 1 compares typical conductivity rates (k) for some common foods. Note that air is a very poor heat conductor; therefore, foods made mostly of air, such as bread and meringue, are also poor conductors.

These concepts provide the basic science that permits us to manipulate the outcome of some classic culinary recipes.

I have chosen three recipes to demonstrate some of the modes of heat transfer I have discussed. In each case, I first outline the cooking method and then explain why certain aspects of the recipe are the way they are from the perspective of heat transfer.

Adapting Ingredients to Cooking Method: Stir-Frying

As we can see from table 1, foods have a range of physical properties that affect the way they cook. Chinese cooks[4] have to take this into account because stir-frying relies on the rapid cooking of many ingredients together in the same wok. Clearly, the wok is very hot, and the cooking times are very short, so by our definition, this is "fast cooking." When temperatures are this high, cooking reactions take place very fast. When cooking is this fast, though, the temperature of the ingredients will probably not be constant throughout; that is, the food will be hotter on the surface than on the inside. This means cooks have to find a subtle balance. Temperatures inside the food, especially in the pork, must be hot enough to kill microbes but not so hot that the structure of the vegetables is irreversibly damaged, resulting in a loss of crunchiness, for example.

The need for speed and heat helps explain two other important aspects of successful stir-frying. First, the sauce is cooked independently of the stir-fried ingredients, which allows the wok temperature to remain very high. If the sauce and vegetables are cooked at the same time, the boiling liquid limits the wok temperature to that of the boil-

3 The space shuttle is able to reenter the earth's atmosphere, withstanding extremely high temperatures, by virtue of its covering of aerogel ceramic tiles, which are similar in structure to highly porous bread. Unfortunately, by the time you read this, the space shuttle fleet will most probably no longer be in service. Bread, by contrast, is still widely available!

4 Strictly speaking, I am referring to Cantonese cooking. It is the closest to what the Western world knows as Chinese cooking. Contrary to popular belief, most of the Chinese population actually eat noodles rather than rice!

ing water, which is not nearly as hot as the temperature oil can reach. Second, it is almost impossible to cook a good stir-fry at home. If you have ever watched Chinese chefs at work, you might recall the huge gas flames they use to keep their woks super hot, much hotter than we typically achieve on a domestic stove top (hob). You might also have noticed that chopstick-wielding chefs keep food constantly moving in these superhot pans.

There is another way Chinese chefs achieve this balancing act when stir-frying: through the precise preparation of their ingredients. Chinese cooking is mostly preparation, followed by a few minutes of frantic wok-based activity. Chefs know by experience that some ingredients cook differently than others, so they cut their ingredients into carefully controlled shapes and sizes (some smaller, some larger) to ensure that all of them are cooked at the same time (in just a minute or two). The science underlying this relates to the relaxation time of a lump of material, which is proportional to the square of its minimum thickness. This means if we double the thickness of a steak, for example, then we quadruple the meat's cooking time—that is, the time needed for the appropriate temperature at the center of the steak to be achieved. In

STIR-FRIED SWEET AND SOUR PORK

For the Pork

2 pork loin steaks
2 pineapple rings
1 onion
1 carrot
Handful of peanuts
3 crushed dried chilies
Peanut (groundnut) oil

For the Sweet and Sour Sauce

Juice from pineapple rings
Light soy sauce
Rice wine
Granulated sugar

Chop the pork, onion, and pineapple into 3-mm-thick pieces; slice the carrot into 2-mm-thick pieces. Heat a wok. Using as little peanut (groundnut) oil as possible, stir-fry the pork, onion, carrot, and peanuts over very high heat for a couple of minutes, until cooked. Remove from the wok to a serving dish. Add the sauce ingredients to the empty wok and cook over medium to high heat until the sugar begins to caramelize and the sauce thickens and darkens (it depends on how much sauce you are making). Pour the sauce over the pork and vegetables and serve with rice.

BARBECUE CHICKEN

Ketchup
Honey
Garlic
Lemon juice
Salt and pepper
2 chicken breasts, thighs, wings, and legs

Mix together the ketchup, honey, garlic, and lemon juice as to develop a marriage of flavors to your liking—sweet, sour, salty, and umami notes are provided by the ingredients listed. You might want to add chili flakes for extra heat. Cover the chicken pieces in the marinade and refrigerate for at least a couple of hours before cooking.* Cook slowly on the barbecue once the flame has died down and the coals are just glowing to produce an even heat. Check the thickest chicken pieces and make sure that the center of the meat is at least 165°F (74°C); the breast usually cooks faster, followed by the thigh. Legs take considerably longer.

---

* The marinating process takes a long time because the flavor has to penetrate into the bulk of the meat. The flavor molecules pass through the structure of the meat by molecular diffusion, which is a very similar process to that of thermal conduction, as described here, except that it is driven by differences in concentration. The analogous property to α (thermal conductivity) is molecular diffusivity.

stir-frying, if the pork were cut into ¼-inch (6 mm) pieces, it would end up undercooked by the time the rest of the ingredients were ready.[5]

Adapting Cooking Methods to Ingredients: Barbecue Chicken

In barbecue cooking, the vast majority of the heat transfer is by means of radiation. The physics of radiation dictate that the amount of energy transferred is proportional to the fourth power of the absolute temperature (that is, in degrees Kelvin, or temperature in Celsius plus 273 degrees), such that the heat transfer rate from coals at 250°C is 15 percent of that at 500°C. This heat radiates, or travels in waves, to the surface of the chicken and then needs to be conducted into the bulk of the meat to achieve a (safe for eating) high core temperature. The rate of conductive heat transfer from the surface of the meat to its core is more or less linearly proportional to the surface temperature. In other words, if the surface temperature doubles, then so does the rate of heat transfer to the meat's interior. Compare this with the heat input rate from the hot coals, which increases much faster than the heat being conducted away from the

---

5 This is also why it is hard to properly cook chicken legs—with their uneven drumstick shape—on a barbecue. The bulbous end requires longer, slower cooking than the rest of the leg.

surface into the bulk of the meat. This leads quickly to very high temperatures gradients, with the consequence that the surface of the meat blackens and burns before the inside is even warm. The situation is exacerbated in frozen meat that is recklessly defrosted and barbecued, because food poisoning then becomes a significant risk. The solution is either to cook the meat over much cooler coals so that the heat supplied has time to distribute inside the meat or, even better, to selectively freeze the surface of the meat (as suggested in chapter 23). Alternatively, the meat can be microwaved to cook the center before finishing it off on the barbecue, or the meat can be cooked *sous vide* before browning it over hot coals.

Adapting Cooking Methods to Ingredients: Frozen Florida

In conventional baked Alaska, a lump of ice cream is coated with a layer of meringue and then the whole thing is cooked in a very hot oven, say 480°F (250°C), for a very short time, about 10 minutes—long enough to color the meringue but not so long that the ice cream completely melts. The ice cream survives almost intact because the meringue conducts heat very inefficiently, insulating the ice cream, and because the ice in the ice cream needs a large amount of energy to melt. In the conventional recipe, heat is supplied from the outside to the entire surface of the dessert and the dish it sits on. This explains the need for a base layer of dry cake, another material with low conductivity. It is there to insulate the ice cream from the heat being conducted to it from the very hot baking dish.

To put this in terms of numbers (and you will have

BAKED ALASKA

8 medium egg whites
3 g cream of tartar
8 ounces (225 g) granulated sugar
¼-inch (0.5 cm) slices of Madeira or another cake
    with a dry texture
1 quart (1 L) vanilla ice cream

Preheat oven to 480°F (250°C). To make the meringue, combine the egg whites, cream of tartar, and half the sugar in a bowl. Whisk for about 3 minutes at medium to high speed. Add the rest of the sugar and whisk for another 4 minutes. A very firm, glossy, and stable meringue will form. Lay the slices of cake in an ovenproof dish. To create an insulating base, add a generous scoop of ice cream and cover with meringue. Place the dish in a very hot oven for about 10 minutes, until meringue is lightly browned. Modern chefs are likely to dispense with the oven and just brown the outside of the meringue using a blowtorch, which is of course cheating!

to trust me here!), we can calculate the heat flux through our meringue layer during the "baking" process. This is done by multiplying the thermal conductivity by the heat-source temperature and the exposed area of the food, and then dividing this by the thickness of the food. Suppose we have a layer of meringue of about ¾-inch (2 cm) thickness over the ice cream. Table 1 tells us that k for meringue is about 0.05 (W/mK), which means that the heat flux is about 0.05 W/mK × 250°C = 0.02 m ≈ 625 W/m². Assuming that the ice cream is approximately a hemisphere with a 4-inch (10 cm) radius, we have a contact area of about 0.06 m², and hence the heat flux is about 35 W. Now, the mass of ice cream is about 1 kg, with a latent heat of about 334 kJ/kg. Therefore, in 1 second, we can melt about 0.01 percent of our ice cream. After cooking the dish 10 minutes at 480°F (250°C), our rough calculation suggests that we would melt about 6 percent of the ice cream. Still, more than 90 percent of the ice cream remains intact under these conditions!

What happens if we try to make a baked Alaska in the microwave? Theory suggests that because microwaves heat the volume rather than the surface of food, the meringue will neither get hot enough before the ice cream melts nor will it brown. However, what if we turned the baked Alaska inside out so that the meringue was on the inside—what then? The meringue, with its very low diffusivity (α), would insulate itself and get very hot in the middle without conducting much heat to the surrounding ice cream. The problem is, the microwaves would simultaneously heat and melt the ice cream layer by directly interacting with its molecules.

One could, however, replace the meringue with, say, fruit puree and make a hot, fruity filling inside a cold ice cream bombe. This was first demonstrated by the late Hungarian physicist Nicholas Kurti in "The Physicist in the Kitchen," a talk given to the Royal Society of London in 1969. In his experiment, he used meringue covered with ice cream with a filling of apricots, sugar syrup, and apricot brandy. Although he did not call it a frozen Florida at the time, he later used the name in an article published in *Scientific American* in 1994. Now, if you make the ice cream very cold, the relaxation time of the water molecules inside becomes much longer, which means that they will absorb less microwave power than the water in the fruit puree. Another thing to consider is that the heat will conduct very rapidly from the hot fruit core into the surrounding ice cream. The secret to making this work is therefore twofold: the ice cream must be supercold and thermally isolated from the hot fruit. And yes—you guessed it—you can achieve this trick by encasing the fruit in slices of Madeira cake.

The way humans understand and control heat has come a long way since the discovery of fire. The concept of cooking, ironically once related to the exclusive *heating* of foods, has also evolved, and now it includes enzymatic reactions, fermentation processes, and acid-induced changes, to name a few. It is gratifying and encouraging to witness the culinary progress that scientists are enabling by coming into the kitchen.

Further Reading

Blumenthal, Heston. 2006. "Pizza." In *In Search of Perfection: Reinventing Kitchen Classics*, 62–84. London: Bloomsbury.
——. 2007. "Baked Alaska." In *Further Adventures in Search of Perfection: Reinventing Kitchen Classics*, 250–274. London: Bloomsbury.
Kurti, Nicholas. 1969. "The Physicist in the Kitchen." A transcript from the Weekly Evening Meeting of the Royal Society, London, 451–467.
Kurti, Nicholas, and Hervé This-Benckhard. 1994. "The Amateur Scientist: The Kitchen as a Lab." *Scientific American*, April, 120–123.

twenty-three

ON SUPERB CRACKLING DUCK SKIN

An Homage to Nicholas Kurti

CHRISTOPHER YOUNG AND NATHAN MYHRVOLD

crispy golden duck skin is a glorious thing—pork crackling, superb. And for many, the best part of roast chicken is also the skin. But cooks face a real dilemma: How can the skin be perfectly crisped without overcooking the meat? The skin must be cooked until dry while the meat stays juicy.

A similar conundrum haunted Nicholas Kurti (1908–1998),[1] a professor of physics at Oxford University who studied ultra-low temperature phenomena. Cooking was a passion of Kurti's and he was troubled by the fact that scientists deemed cooking unworthy of serious study. His book *But the Crackling Is Superb: An Anthology on Food and Drink by Fellows and Foreign Members of the Royal Society* is as much an inspiration to us as it is to the editors of *Kitchen as Laboratory*. Our chapter, an homage to Kurti, purposefully combines elements of his legacy: crispy, crackling skin and low temperature physics. Here we explain how it is possible, with some extra effort, to get great skin and keep the meat tender and succulent by using an approach we believe would have pleased Nicholas Kurti.

1 One of Nicholas Kurti's most celebrated achievements involved nuclear cooling experiments that came within a millionth of a degree of absolute zero, where atoms will stand still (−459°F [−273°C]). To make sense of this temperature, the coldest temperature ever recorded is −129°F (−89°C). It goes without saying that absolute zero must be very, very cold.

Although the science and technology of meat cookery is well-covered territory, skin is something of an afterthought. It should not be. From a biological perspective, skin is a vital organ. This complex material has myriad roles in life—protecting, sensing, regulating, sweating, and absorbing, among them. But when cooking it, you can ignore most of these biological nuances and focus on three of skin's key components.

The first is the mesh of connective tissue, woven mostly from collagen that gives skin strength and elasticity. Skin is tough because of the long helical collagen fibers that give it structure. A collagen fiber is strong and unyielding, but nature weaves individual fibers together for greater strength and to allow some stretch. Think of skin as a softer and more pliable version of fiberglass cloth.

The second important component of skin is water. Skin is surprisingly wet: about 70 to 80 percent water by weight. Thus, skin, like many other plant and animal foods, can be thought of as simply water filled with impurities! The reason that skin does not appear wet is because the water in skin is not free flowing; it is trapped by skin's third key constituent: soft, elastic gel.

The gel itself is delicate and would be torn easily were it not for the fabric of collagen fibers woven within it. This collagen gives it strength to survive the rigors of life. Indeed, biologists believe that the collagen strands actually rearrange their weave at the edges of a cut to stop the mesh from fraying.

The enormous quantity of water bound within skin is essential for the overall health and functioning of a living animal. After all, one job skin does is to prevent the organism it envelops from drying out. But, for the cook in search of crispy skin, this water is a nuisance that must go—eventually. What cooks may not realize is that the water in skin is needed in order to cook the skin in a way that makes it pleasant to eat.

Softening the Skin

Because collagen makes skin tough, tenderizing it requires heating to temperatures high enough to unravel helical collagen fibers. But before this happens, the collagen fibers will contract, shrinking the skin. If you pay careful attention, you will notice that skin shrinks more in one direction than another. This is because the woven mesh of collagen fibers is oriented in a particular direction. As a result, skin is more elastic in one direction than another, which is to say, skin can stretch more lengthwise than widthwise, relative to the direction that muscle fibers elongate. During cooking, skin contracts more in this stretchy direction.

This detail is worth keeping in mind if you want to ensure that a cooked duck breast, for example, ends up fully covered with a crispy skin. You should leave extra skin everywhere around the breast during butchering but then leave more at the top and bottom because the skin will contract the most this way.

As the skin cooks, the corkscrew-shaped collagen molecules will contract, visibly shrinking the skin. Continue cooking the shrunken skin for long enough in a very moist environment, and the collagen molecules will unwind and split into shorter fragments. Scientifically, this transformation is known as hydrolysis—literally, "water splitting"—but cooks know it better as gelatinization, because gelatin is the result. Keep in mind that this is an entirely different process than starch gelatinization (described in chapter 8).

Gelatinization of collagen occurs ever slower as the cooking temperature falls to less than 158°F (70°C), and it increases rapidly as the cooking temperature rises above this threshold. This point is a very important one: converting tough collagen into tender gelatin takes heat, a lot of moisture, and time. Skin starts out with more than enough water for gelatinization to occur, and it can be softened quickly against the high heat of a frying pan, or it can literally take days at the low cooking temperatures of *sous vide* cooking favored by many chefs. Indeed, the *sous vide* cooking temperatures considered ideal for many tender cuts of meat, like duck, chicken breast, and pork loin, are often so low that skin never becomes tender. This problem is one reason why *sous vide*–cooked tender meats are rarely served with the skin on.

Assume for a moment that you have gotten the skin hot enough for long enough for it to become delicate and sticky—a telltale sign that the collagen has been sufficiently gelatinized; you should shift your focus from tenderizing to rendering any excess fat under the skin before you worry about drying the skin to make it crispy.

Getting the Fat Out

Rendering fat might seem like a simple task with a simple solution: just make the fat hot enough that it melts. And at first glance, the layer of fat between skin and muscle looks uncomplicated. Why, you might wonder, is the fat even there? This layer provides cushioning and, more important, insulation for the animal's internal tissues. So it can be thick (as in ducks, for example) or thin (more typical of chickens), depending on how much insulation against heat loss the animal needs. Thickness also varies from place to place on the body. The belly, for example, tends to be generously endowed with fat. This tendency seems to be the result of a set of genes common to all animals with spines. It can be clearly seen in our species by visiting a crowded beach.

All this fat resides within a highly structured and complex tissue composed of many fat-filled cells, which are held in place by a fine network of collagen fibers. For cooks, the salient point is that the layer of fat under the skin does not exist as pure globules that melt easily. Rendering the fat out of this tissue requires rupturing cells and weakening the collagen mesh, which takes the same high temperatures necessary to soften skin. The problem is that fat-rendering temperatures are higher than desirable for cooking tender cuts of meat. Another complication is that rendered oil escapes through cracks and fissures in the skin, whereas the oil is trapped in intact skin. That is why many cooks score the skin of fatty meats, like duck breast.

Another good tactic is to use the bristles of a stiff wire brush (such as an [unused!] pet brush) to poke holes through the skin and into the fat. This simple maneuver provides a lot of channels that allow oil to escape. Because the holes are so tiny, it leaves the skin largely intact, looking undisturbed and protecting the flesh underneath from overcooking. When rendering, do not remove all of the fat between the skin and flesh. Leave behind a thin layer. After cooking, this layer acts as a moisture barrier, which keeps skin from quickly wicking juices from the meat and becoming soggy.

Puffing Makes Skin Extra Crispy

Although straightforward drying works, we recommend a different system for getting the water out of cooked skin. It is tricky, but when it is done just right, the skin will puff into a delicate structure that crisply shatters when bitten. The first step follows the conventional approach—driving the water out of the gelatinized skin with heat—but the skin should not be allowed to dry out completely.

Actually, it is difficult to make the skin too dry at this stage; often it is difficult to get the skin dry enough. That is because skin absorbs juices from the meat under it. One way to address this problem is to cook the meat until it, too, becomes dry. The skin will then be crunchy, but the meat will have been overcooked. A better way to get the skin dry enough is to physically separate it from the underlying flesh. One approach is to cook the meat and skin separately and then recombine them on the plate. Even just loosening the skin from a roast makes a profound difference in the drying process. This trick is used for Peking duck—the intact skin is inflated until it tears away from the flesh—and it works well for other poultry, too. While the meat cooks, steam and hot air hold the loosened skin away from the moist flesh so that the skin can dry. Skin treated this way tends to stay crisp because, even after cooking, ongoing evaporation holds the dry, glassy skin away from the moist flesh, permitting the skin to more slowly absorb escaping juices.

No matter how dry the skin may appear, there will always be a little bit of water trapped within it. This is largely because a significant fraction of water in skin is tightly bound to the surface of the proteins that constitute skin. Evaporation will remove free water from skin, but the bound, or vicinal, water is slow and difficult to vaporize. Not to worry; trapping a little water inside the skin is a good thing. Indeed, it is crucial for getting the skin to puff up.

Applying intense heat quickly transforms tiny liquid water droplets into expanding bubbles of steam, and—poof—the skin puffs.

For cooking, a very hot oven will do, but a deep fryer or a pan with a generous layer of oil is even better. The heat of the hot oil makes the glassy skin soften and become rubbery, just as heating hard caramel makes it pliable. The cooking has to happen fast, though, because the trapped water begins to vaporize almost immediately upon heating.

If the glassy skin is hot enough to become soft, which requires heating the dried skin to between 355 and 390°F (180–200°C), the expanding steam-filled bubbles will stretch the skin into a delicate cellular structure. If the skin is not heated enough or is heated too slowly to become soft and elastic when these water droplets vaporize, the steam simply fractures the skin and escapes without puffing it up.

Whether the skin puffs depends also on the number of water droplets present. Too many will prevent the skin from getting hot enough fast enough, because more than just boiling away the water needs to occur: the steam must be superheated to temperatures of more than 212°F (100°C). And that will not happen if there is a lot of liquid water around. But if there are too few droplets, there will be nothing to make steam from.

In practice, drying the skin until it is just moist enough to puff takes a bit of trial and error. If everything goes right, the skin will be riddled with fragile bubbles that easily shatter in the mouth, providing a distinctly crisp sensation.

Unfortunately, the high cooking temperatures needed to dry and puff the skin will overcook the flesh of tender cuts of meat, like duck breast or pork loin. Some people view this as merely an unavoidable trade-off: you can have either crispy skin or succulent flesh—take your pick. We disagree, because we have a solution that makes it possible to have both at the same time: freezing.

Cryocooking: A Novel Approach

It takes a surprisingly great amount of heat to melt ice, and until it melts, the temperature can never be above freezing. Our technique of cryocooking takes advantage of this simple fact to crisp the skin without overcooking the tender flesh it covers. We apply this method to the cooking of a duck breast, but the technique can easily be adapted to many other meats and seafood. Because a duck breast is covered by a particularly thick layer of fat, we need two distinct steps: a cryorendering step done well before service and a cryosearing step done as part of the service.

Begin by butchering the duck breast so that it is surrounded by extra skin. Fold the extra skin up and over the backside of the breast (figure 39). This compensates for the shrinking of the skin during cooking, leaving a crisp, even layer covering the flesh after the final searing. Folding the skin over the breast also protects the flesh from escaping steam and splattering oil during searing. Next, perforate the duck skin and underlying fatty tissue (figure 40). As mentioned earlier, we recommend a wire pet brush because it leaves no visible marks after cooking.

Start the cryorendering step by freezing a layer of the flesh underneath the skin. As the skin cooks and the fat renders, the icy flesh will thaw slowly, and until it does, heat cannot penetrate further into the duck breast. The best way to do the freezing is skin side down on a flat sheet of dry ice. At nearly −112°F (−80°C), dry ice will freeze the skin, fat, and flesh rapidly. This damages the flesh as little as possible and makes the frozen layer very cold, which forestalls thawing of the

Figure 39 The duck breast is carved so that extra skin can be used to keep escaping steam and splattering oil from cooking the flesh as the front is seared.

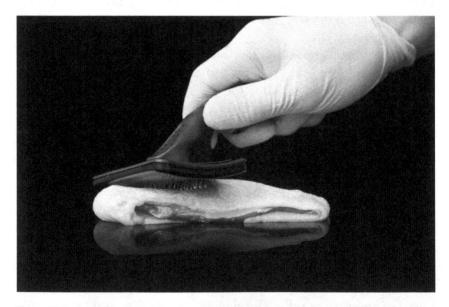

Figure 40 Perforate the skin to allow the oil to escape. A wire pet brush is a good tool for making pricks in the skin and underlying fatty tissue that are so small they become invisible after cooking.

Figure 41 The breast is placed skin side down on a block of dry ice for about 25 minutes. About a quarter of the flesh will freeze. A weight is placed on top to flatten the breast against the ice (and later the pan or griddle). A muslin satchel filled with loose pieces of metal, which conforms to the contours of the meat, makes an ideal weight.

flesh. Keep the breast pressed flat against the sheet of dry ice using a satchel of loose ballast (figure 41); once frozen, the breast will become too rigid to flatten against a hot griddle for even searing. Keep the duck on the dry ice for about 20 minutes, which should freeze a little less than one-quarter of the flesh underneath the skin.

Finish the cryorendering by searing the frozen duck breast in a pan or on a hot griddle for about 5 minutes. Again, keep the skin pressed flat. The thin skin and fat will thaw quickly, but the thicker layer of icy frozen flesh will thaw slowly, which prevents the meat from cooking. This step will soften the skin and render some of the fat from underneath the skin. But do not render out all the fat at this stage. More will be rendered during the final crisping step, and you want the finished breast to end up with a thin layer of flavorful and water-proofing fat between the crisp skin and juicy flesh.

At this point, the skin needs to be dried and the flesh cooked. The cooking temperature does not need to be any hotter than that suitable for the tender flesh of the duck breast. This is because the collagen in the skin has already been substantially gelatinized and the fat partially rendered out by the cryorendering step.

A basic oven will work, but be aware that water evaporating from the skin cools the breast, lowering the temperature at which the meat actually cooks. If you want the duck cooked to an internal temperature of 136°F (58°C; our personal preference), then the oven temperature needs to be set somewhat higher to allow for this evaporative cooling.[2] How much higher depends on the rate of evaporation from the meat, which is contingent on the humidity inside the oven. It is hard to predict.

Combination, or combi, ovens—convection ovens that steam, poach, roast, broil, and bake—and controlled-vapor ovens are ideal because they eliminate this guesswork by controlling the temperature and the humidity. This means that they control how much evaporative cooling occurs during cooking. If you are using a combi oven, set the temperature to 158°F (70°C) and the humidity to 50 percent. These settings will eventually cook the flesh to a temperature of 136°F (58°C), but they will also ensure that the skin dries and puffs to a crispy texture later. For a controlled vapor oven, set the doneness to the desired core temperature and then set the browning 20°F (10°C) higher than this temperature for the same reasons.

Admittedly, this is all a bit tedious, and it is tempting to simply try cooking the duck breast *sous vide* in a water bath and be done with it. However, this should be avoided at all costs! Contrary to popular opinion, the inside of a *sous vide* bag is not a dry cooking environment. Evaporating water quickly saturates the inside of a *sous vide* bag with moisture, and the duck skin will not dry out.

Once the duck breast has been cooked through and the skin dried, the duck is finished with the cryosearing step. Return the cooked duck breast to the dry ice. This might seem like overkill in cooling the dried skin down before puffing it, but actually, a thin layer of the

---

2 When it comes to cooking meat and seafood, commonly prescribed cooking temperatures almost always result in *over*doneness! It is often claimed, for example, that you must cook meat such as duck breast to an internal cooking temperature of 145°F (63°C) to prevent food-borne illnesses. This statement is totally false. The Food and Drug Administration requires *no* specific internal temperature for duck, or other meats for that matter.

If you study the microbiology at work, you will learn that there is very little need to prescribe a specific internal temperature because the inside of a healthy muscle is sterile. The animal's immune system took care of eliminating any pathogens in the muscle. (If it had not, the animal would not be healthy.) So unless the meat has been cut or punctured, the interior will remain sterile.

It is the surface of the meat that you need to worry about because handling it can spread bacteria and make you ill. In the case of our cryocooked duck breast, some puncturing of the flesh occurs when the skin is perforated. This process can, potentially, carry bacteria deeper into the meat. When the breast is seared, however, the high temperature quickly kills any bacteria that have taken up residence on, or just beneath, the surface and thus ensures the dish is safe to eat.

cooked flesh needs to be frozen to keep it from overcooking while the skin is fried and puffed.

Because the puffing happens quickly, the frozen layer of flesh should be thin; about 3 to 5 minutes on the dry ice will do. Again, weight the breast down flat: a warm satchel of loose ballast heated in the oven with the duck is useful for this. Not only will it keep the duck skin pressed flat against the dry ice, but it will keep the cooked flesh above from cooling.

Finish the breast by searing the skin in a hot pan filled with enough oil to completely submerge the skin. The skin should quickly crisp as it puffs and turns a golden brown. Remove the duck breast, blot off any excess oil, trim the extra skin that is still folded over the flesh, season, and serve.

Further Reading

Kurti, Nicholas, and Giana Kurti, eds. 1988. *But the Crackling Is Superb: An Anthology on Food and Drink by Fellows and Foreign Members of the Royal Society.* Bristol: Institute of Physics.

Lepetit, J. 2008. "Collagen Contribution to Meat Toughness." *Meat Science* 80:960–967.

Myhrvold, Nathan, Chris Young, and Maxim Bilet. 2011. "Meat and Seafood." In *Modernist Cuisine: The Art and Science of Cooking*, vol. 3, *Animals and Plants*, 116–150. Seattle: Cooking Lab.

Roudaut, G., C. Dacremonta, B. Vallès Pàmies, B. Colasa, and M. Le Meste. 2002. "Crispness: A Critical Review on Sensory and Material Science Approaches." *Trends in Food Science and Technology* 13:6–7.

Steingarten, Jeffrey. 2002. "As the Spit Turns." In *It Must Have Been Something I Ate: The Return of the Man Who Ate Everything*, 36–43. New York: Knopf.

twenty-four

SWEET PHYSICS

Sugar, Sugar Blends, and Sugar Glasses

NATALIE RUSS AND THOMAS VILGIS

table sugar is ubiquitously used in desserts, baked goods, drinks, and countless other food items, mainly to make them sweet. However, sugar shows properties—not easily visible—that go much beyond sweetness. For example, sugar molecules bind water molecules in a "hydrate shell"; in so doing, they increase the viscosity, especially at high concentrations, of the liquids that contain them. Also, sugar depresses the freezing point of water, a property that is exploited in the making of ice cream (described in chapter 17). In simple terms, sugar serves many more purposes in the kitchen than the ones we know best.

One purpose is the making of sugar glass. This elaborate ornamental sugar showpiece is most commonly seen in culinary competitions and at fancy banquets. The talented pastry chefs spend hours, if not days, pulling and blowing sugar, as their complex artistic visions gradually materialize, taking the form of fragile, elaborate sculptures. This pulling and blowing process resembles the work of glassblowers; it transforms the solid sugar into what appears to be fine glass. Other sugar-based creations, which are perhaps more familiar, also play on the transformation of sugar from the granulated and solid crystalline state to the glassy state. Cotton candy—one of those treats loaded with nostalgia that takes you right back to your childhood—is the ultimate sugar glass. Cotton-candy making relies on the fact that sugar can show liquid-like behavior within a relatively wide and high temperature window. As the sugar is deposited inside the hot and fast-spinning chamber of the cotton candy machine, it melts. The liquid sugar is then spun out of the chamber in the form of a continuous, fine edible thread. Cotton candy possesses expansive surface areas that facilitate the uptake and delivery of all sorts of aromas and flavors and that cause the sudden sensation of liquefaction in the mouth. This property has made cotton candy a popular treat even in the context of high-level gastronomy.

Let us take a closer look at the physical chemistry that makes sugar glass possible. Consider two basic features of sugar. First is its great deformability and its large changes in flow behavior as it melts over a wide temperature range. The second is related to the thus far unknown ability of sugar glasses to adsorb

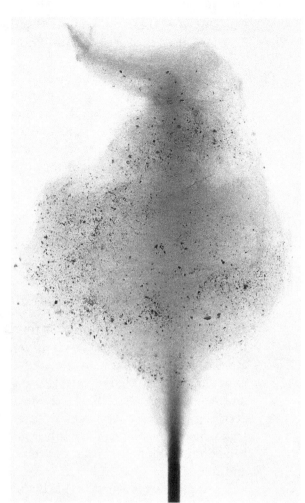

Figure 42 A fine cotton candy aromatized with cubeb pepper combines sweet and hot notes on the tongue. The sudden dissolution of the glassy sugar in the mouth and the explosive flavor release yield unusual taste sensations. (Photograph by Peter Schulte; reproduced with kind permission from Jörg Sackmann, *Aromen* [Wiesbaden: Tre Torri, 2008])

volatile aromatic compounds. To illustrate this, figure 42 shows cotton candy that has been aromatized with cubeb pepper. The culinary effect is as complex as it is sensational: sugar dissolves very quickly in the mouth, releasing the pepper's many aromas. Almost simultaneously, the individual pepper pieces provide the sensation of heat, while emitting their complex spicy fragrance and sensations. However, something else happens: during the dissolution of the sugar filaments, the aroma compounds absorbed within seem to make their way into the retronasal region. This allows a slightly hot taste to linger in the mouth.

Ideally and aside from the aroma and taste-release considerations, once the cotton candy is made, it must "stay alive" for an extended period of time. This means that it should not collapse by absorbing the natural humidity of the air. The initial invention of cotton candy required some thinking about the nature of the fine sugar filaments. In particular, the sensitivity of the sugar to water had to be reduced. This leads us to two key questions: What is hidden behind these observations and developments? How can we use what is hidden to design new culinary creations?

What Can We See?

An examination into the structure of sugar before and after melting and following cooling of the melt reveals a lot about sugar's chemistry. Granulated sugar has a *crystalline* structure. The sugar molecules are very much ordered with respect to one another, not unlike bricks in a wall. At the opposite end is the *amorphous* state, in which the molecules, randomly packed together, have no organization whatsoever. Sugar can exist as a solid in both the crystalline and amorphous states. Scientists refer to solid, amorphous sugar as a glass, because its behavior is analogous to that of regular window glass—hard, brittle, and fragile.

Sugar glasses are usually made by first melting crystalline sugar and then quickly cooling down the melt. Molten, liquid-like sugars are clear and transparent amorphous liquids that, as long as their temperatures do not exceed the caramelization point, will remain clear upon cooling. It is during cooling that a dramatic change in viscosity occurs. As the temperature decreases, the liquid becomes increasingly more viscous, until it cannot flow any more. At this point, the sugar remains very soft and flexible; with further cooling, it solidifies. It is this shift from a very soft and flexible "rubber" into a hard and brittle "glass" that polymer scientists refer to as the glass transition temperature (T_g). The temperature range in which the sugar is below the melting and above the glass transition temperature enables the creation of cotton candy and is of the utmost importance to pastry chefs—this is when they can pull and blow sugar.

Based on several experiments and tastings, we have reason to believe that the glassy state enhances the culinary sensation in a very pronounced way, far beyond sugar's sweetness. In this context, although we have spoken only about table sugar, many other sugars show the same features; that is, they all have a glass transition. They also have small but essential differences that can make them uniquely useful for particular recipes.

Table sugar, or sucrose (saccharose), has a defined melting point, well-understood flow properties, and quickly undergoes browning or caramelization reactions upon heating. It is also, for some applications, a little too sweet. It might seem counterintuitive, however, but for many culinary applications, either in the avant-garde kitchen or in industrial food design, the aforementioned properties of sugar are actually a nuisance. Other sugars, such as isomalt and trehalose, do not show fast browning during heating, are considerably less sweet, and can better take up other tastes and aromas without overpowering them with their sweetness.

To better understand sugar glasses, we suggest the simplest of experiments: take a spoonful of table sugar, which we now know is in the crystalline state; heat it very carefully in a pan so as to avoid caramelization; and when it has completely melted, pour it on a cool plate and observe what happens. The cooled sugar will remain as transparent as a window—it has transformed itself into a sugar "glass." Next, take this "glass" of sugar and try dissolving it in water. You will quickly realize that "glassy" sugar dissolves much faster than crystalline sugar (for pieces of similar size). For the same reason, a piece of glassy sugar will appear to deliver sweetness on the tongue more quickly than its crystalline counterpart. Seeing how we can manipulate sugar's chemistry, it is tempting to design sugar blends with customized properties for special culinary purposes, such as a well-defined glass transition temperature, to extend the time and the temperature range in which the sugar blend can be easily pulled or blown, but also to better control its stability. The idea is then to mix different kinds of sugar with appropriate (low) sweetnesses and relatively different glass transition temperatures. Such sugar blends will show a glass transition temperature somewhere between those of the two pure components and a lower sweetness profile. This would in principle allow us to aromatize the glass without a dominating sweetness masking the flavor we intend to deliver.

What is of scientific importance when performing these experiments in the kitchen? What is important to consider when blending sugars? It should be merely a matter of mixing them at a certain ratio, melting the blend, and then cooling it. Simple? Let us look further into this process—from a physicist's point of view.

Sugar Blends

The simplest question in this context is: If the two sugars to be blended have two different glass transition temperatures, what is the overall glass transition temperature of the mixture? This question makes sense only if we know for sure that they mix at the molecular level once we melt them together. The glass transition temperature is very much dependent on the sugar-blend composition, and the ideal blend is determined by use of the so-called Gordon-Taylor equation, which was originally proposed for polymer blends. We will not go into the mathematical details and other complicated matters; we prefer to focus on the practical aspects of sugar blending.

As an example, we look at the two disaccharides sucrose and trehalose, which differ in their molecular structure and, hence, show major differences in their physicochemical properties. Sucrose consists of one glucose and one fructose molecule, whereas trehalose is composed of two glucose molecules—this is why they are called disaccharides: they each possess two sugar molecules. Their glass transition temperatures and melting temperatures, measured through a technique called differential scanning calorimetry, turn out to be significantly different: the glass transition temperature is approximately 131°F (55°C) and 270°F (133°C) and the melting

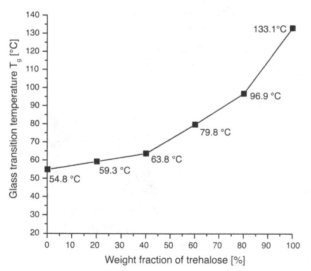

Figure 43 The change in the glass transition temperature of sucrose–trehalose blends as measured with the aid of a differential scanning calorimeter. The glass transition temperature changes continuously with trehalose concentration. Therefore, a precise design for gastronomic applications is possible by adjusting sweetness, flow behavior, and solidification temperature by blending.

temperature is about 367°F (186°C) and 397°F (203°C) for sucrose and trehalose, respectively. Moreover, trehalose has the advantage over sucrose in that it is roughly half as sweet and does not undergo caramelization at the so-called soft- or hard-ball temperatures used in candy making. For reference, we present only the results for the glass transition temperature for the sucrose–trehalose blends (figure 43).

Blending, Tempering, Kinetics

To get perfect culinary results, the mixing has to be perfect, down to the molecular level. Molecularly speaking, "perfect" mixing means that the mixture will have to show a single glass transition temperature, which reflects the true miscibility—like water and alcohol—of the two sugars. In culinary terms, perfection requires that the flow behavior of the blend be similar to that of the pure components. Since molten sugar moves slowly—think of honey, for example—it is necessary to give the blend time to mix homogeneously. A simple yet relevant experiment exquisitely illustrates this. In a coffee grinder, pulverize first the sucrose and then the trehalose. Blend them together at a 1:1 ratio in a Pyrex cup. Proceed to slowly heat up the mixture in an oil bath set to a temperature of 420°F (220°C). The blend will melt, but you will observe how sucrose melts first, followed by trehalose. A sample taken immediately after the melting of the sucrose is shown in figure 44.

We found that the flow behavior of sucrose–trehalose mixtures depends significantly on the tempering and stirring time. This manifests itself in the creation of large "pearls" during the development of the sugar filaments, as can be seen in figure 45. This indicates that on the molecular level, the sugars are not homogeneously mixed. In such a nonhomogeneous state, the blend cannot be pulled or blown in a reliable and controlled manner. Even after holding the blend for

Figure 44 Sugar filament of a 50:50 sucrose–trehalose mixture drawn immediately after melting. The filament is not homogeneous, and the lumps are showing at almost equal and characteristic distances.

Figure 45 Sugar filaments of a 50:50 sucrose–trehalose blend after different tempering times. The two pure-component filaments are shown: (*farthest left*) sucrose and (*farthest right*) trehalose. The three other filaments represent the blend right after melting: 10 minutes and 15 minutes tempering time (*left to right*). The sucrose content is responsible for the light caramelization.

10 minutes at a temperature in between the melting temperatures of the two sugars, the blend will not homogeneously mix. Pearls and lumps are less pronounced but still present. Only after 15 minutes can the mixture be deformed and regular fibers spun. Hence, the conclusion is very simple and general: sugars can be blended and their mechanical and flow properties designed by manipulating their concentrations and tempering times.

Now let us explore how the nature of the solid state (glassy or crystalline) influences the aromatization of sugars. This is a widely open and unexplored field and very relevant for many culinary applications. Again, this is another situation in which the fundamental questions in physics and chemistry meet at the point of flavor, in which visible length scales are addressed down to the size of molecules. A prominent example is found in chocolate. The glassy sugars used in chocolate may allow for the uptake of more aroma compounds than do crystalline sugars. Because of the disordered nature of the amorphous state, the adsorption and diffusion of aroma compounds on amorphous sugar surfaces is hypothesized to be much better than that of crystalline sugars, in which a highly ordered structure hinders the ability of the sugar to interact with its environment.

How can this be visualized? Roughly speaking, a sugar crystal has a highly regular and ordered surface. This gives it a very well-defined adsorption and desorption energy in relation to aroma compounds, which serves to restrict their interaction. Whereas the amorphous structure of sugar in a glassy state allows the aroma compounds to latch on to edges, wedges, and other irregular topographical features. Such irregularities create more opportunities for the aroma compounds to make contact with the surface. This multifaceted interaction between aroma compounds and sugar surfaces is essential to achieving a higher concentration of flavors in any finished dish.

Why Should the Glass Transition Temperature Be a Culinary Parameter?

In the presence of molten liquid sugar, small droplets of aroma compounds (which are mostly hydrophobic) reside at the surface. As we stir the sugar blend, we help emulsify, or disperse, these droplets within the molten sugar. Physics tells us that these emulsified droplets generally will want to associate with one another to form larger droplets and minimize their surface energy. However, as the blend cools, its viscosity increases and, eventually, it solidifies into a glass that contains droplets trapped in a random environment, and these droplets cannot escape from within the glassy sugar matrix. The glassy sugar thus ends up being highly aromatized with hydrophobic aroma compounds. Knowledge of the sugar's transition temperature plays an essential role in the design of these systems: the lower the sugar's transition temperature, the better the aromatization obtained because more time is allowed for mixing. Cooler temperatures also mean a smaller loss of volatiles.

So, what is the best sugar for making glasses, keeping in mind that they have to keep their shape and properties under humid and hot conditions, such as those in a restaurant? We know that the transition temperature of the blend should be as high as possible. A sufficiently high transition temperature ensures that there will be low-motion sugar molecules in the amorphous matrix at room temperature and thus a lower water solubility. This means that the water molecules dissolved in the air cannot easily penetrate the glass and soften it. Adding increasing amounts of trehalose to sucrose thus helps stabilize the candy (see figure 43). The sugar must have a balanced transition temperature to enable enough aromatization with the maximum stability.

What Is This Good For?

It is good for gastronomy, of course. Take, for example, a blend of not very sweet sugars, such as isomalt, lactose, and trehalose. Mix in finely ground salt, aromatize with very dry lemon powder, and melt the blend as described earlier. Once melted, dip the tip of a spoon into the molten blend and pull it slowly upward to form a thread or filament. The thread will soon solidify (sooner with increasing trehalose content). You can then add this filament to steamed sea fish as a crispy component. This, of course, is also possible with common table sugar, but table sugar will more often than not caramelize, which will impart a flavor you might not want to pair with fish.

Another example is the now-famous olive oil bonbon served at José Andrés's Minibar in Washington, D.C. Approximately 3 ounces (100 g) of crystalline isomalt, a disaccharide composed of one glucose molecule and one mannitol molecule, is melted so as to render it a viscous liquid. Dipped into the melted isomalt is a ½-inch (1.25 cm) plastic ring (wide enough to avoid burned fingers). It is quickly pulled out from the pot, causing a film of isomalt to form at the bottom of the ring. It is cooled for about 10 seconds to allow the isomalt to thicken and, immediately, with the help of a spoon, 3 milliliters or so of olive oil are poured from the opposite end of the ring. The oil pulls down the film of isomalt, which, as it cools, will form a solid, amorphous, drop-shaped casing with a long tail that contains the liquid olive oil. The bottom of the drop is lightly moistened and dipped into a mix of Maldon salt and powdered vinegar.[1]

We never know who the sous-chef really is then: physics or chemistry. It does not matter—what matters is the chef's taste and the incorporation of science as a deliciously smart spice.

Further Reading

Cammenga, H. K., K. Hopp, K. Gehrich, and G. Ziegleder. 2008. "Über das Adsorptionsverhalten bei amorphen Zuckern." *Lebensmitteltechnik* 40:54–59.

1 For an image, visit http://cltampa.com/dailyloaf/archives/2010/10/22/foodies-eat-your-hearts-out-epicurean-adventure-at-jose-andres-minibar-dc.

Seo, J.-A., J. Oh, H. K. Kim, Y. H. Hwang, Y. S. Yang, and S. J. Kim. 2005. "A Study of Glass Transition Temperatures in Sugar Mixtures." *Journal of the Korean Physical Society* 46:606–611.

Vilgis, Thomas. 2010. *Das Molekül-Menü: Molekulares Wissen für kreative Köche*. Stuttgart: Hirzel.

twenty-five

COFFEE, PLEASE, BUT NO BITTERS

JAN GROENEWOLD AND EKE MARIËN

in 2004, we decided to organize small-scale courses on molecular gastronomy for cooking enthusiasts. In addition to being a professional caterer, one of us was teaching many cooking classes—the perfect venue for gleaning what is of particular interest to the enthusiastic home cook. There were questions like, Why does a béarnaise split? What is the logic behind prescribed oven temperatures? Each time a theoretical topic was treated by the "chemist" (Groenewold), the "cook" (Mariën) asked about its practical use: What can you do with it? What kind of story would be relevant to someone actually cooking it? A few years later, the fruits of our collaboration, *Cook & Chemist*, culminated in the publication of a second book, both titles putting the principles of molecular gastronomy into practice.

One of the recipes that we developed in this process was an orange ice cream. We added flavor by means of orange peels. In order to avoid bitter flavor, the orange peels were extracted in butter, and this butter was added to the ice cream composition. The resulting ice cream had a strong and pleasant orange flavor. A variation on that recipe is a coffee ice cream. The recipe also aims at avoiding bitterness and is given at the end of this chapter.

At times, however, in both ice creams, the infused butter imparted some bitter notes, suggesting that the butter sometimes also extracts bitter flavors from the orange peel or the coffee. Frustrating, perhaps, for a chef, but these culinary missteps are often inspirational for the scientifically inclined. Indeed, this observation was a good excuse to dive into the details of solvent extraction and to come up with a few concepts and rules that may be relevant to any molecular-gastronomy aficionado.

Solvent extraction is a technique well known among chemists and chemical engineers. It is used as a means to separate specific compounds from a mixture. Historically, this technique is used in the perfume industry and is known as enfleurage. Delicate aromas from flower petals are extracted at low temperature in animal fat. The fat is subsequently molten and mixed with ethanol. The ethanol is then separated from the fat, and from it the desired flower aroma is extracted.

The principle is based on the different affinities that any given molecule has for different solvents. For example, a perfume or a flavor molecule will have affinities for water and vegetable oil, which are usually quite different. The technique of solvent extraction is classically already used in the kitchen—for instance, in the preparation of lobster butter. In his book *Sauces: Classical and Contemporary Sauce Making*, James Peterson includes a recipe for lobster butter. By crushing lobster shells and subsequently heating them with a mixture of butter and water, flavors from the shells are extracted that are hard to obtain otherwise. The idea is to deploy butter to obtain oil-soluble pigment and flavor from the shells.

In enfleurage, the contact between fat and the flower petals is direct. Likewise, in the preparation of lobster butter, there is direct contact between the lobster shells and the fat–water mixture. The water is added because it weakens the lobster shells and makes it easier to scoop off the fat. The transfer of aromatic compounds can, however, occur indirectly. For instance, in the kitchen, a type of gas-phase extraction also is classically known. If an odorous truffle is sealed in a container together with some butter, the butter will eventually absorb the aroma from the truffle. Instead of butter, eggs can be used; it is the fatty content of the yolk that absorbs the truffle fragrance. Some people do this out of respect for the truffle because it is bound to lose its aroma eventually, even in a sealed container; therefore, it is a way to maximally use the truffle flavor. However, it must be noted that the presence of the butter (or eggs) accelerates the speed with which the truffle loses its odor. This is a function of the huge capacity of fats to accumulate odor molecules.

Solvent extraction techniques can be used to infuse cheeses with flavors. Instead of the butter, use Camembert or Gouda, and seal them up with, for instance, chopped onions, mushrooms, truffles, and herbs, as discussed by Pierre Gagnaire and Hervé This (2005). Also, the same concept can be applied to the preparation of pâté. In fact, it is a shame not to infuse the fat in the pâté with other flavors, as the fat is used in the pâté anyway. We can do so by gently cooking the pork fat together with all kinds of aromatics—apples, onions, and herbs—before actually preparing the meat batter.

The person who brought the idea of solvent extraction to our attention was Hervé This (1995b). He described a way to "multiply all flavors by a factor of two." Take a jar and pour some oil and water into it. Then, an ingredient—such as vanilla seeds—can be added to the water-and-oil mixture. After shaking and a subsequent wait, the oil-soluble compounds can be separated from the water-soluble ones by separating the water and oil phases. Shaking has the effect of increasing the contact area between oil and water, which promotes the transfer of flavor to the oil phase. The oil and water phases will have distinctly different flavor profiles. By this method, each existing flavor can therefore be separated into two distinct ones: the water-soluble flavors and the oil-soluble flavors. In principle, this can be done with any kind of ingredient (as long as the ingredient does not contain too many surface-active compounds, that is; but there will be more about this later). Obviously, it is not just molecules with flavor that can be divided between oil and water but also other types of molecules, such as color compounds. Red beets present a nice example. The red coloring agent (betanine) from the beets is water soluble and the oil remains virtually colorless. This oil has an interesting earthy flavor that is dominated by the compound geosmin. After the mixing procedure described earlier, a cook may use both the water phase with its color extract and the oil phase containing the flavors. For instance, with the oil, a white beet mayonnaise may be prepared and the remaining beet juice can be used in a Bloody Mary (cocktail) for extra color and sweetness without incorporating the earthiness of the beet.

It is hard to predict the kinds of flavor profiles that can be created by solvent extraction, but a few basic rules apply. It turns out that the basic tastes as perceived on the tongue (sweet, sour, salt, bitter, and umami) are generally water soluble. The logic here is that good water solubility facilitates the tastants reaching our taste buds; water-soluble molecules can penetrate the saliva on the tongue more easily than molecules that dissolve poorly in water. The oil-soluble compounds are predominantly volatile aroma molecules. It is a good rule of thumb to remember that volatile aroma compounds are usually oil soluble and that taste compounds are water soluble. Note, however, that there will always be exceptions. Indeed, many oil-soluble compounds have a bitter taste and some aroma compounds have a preference for water.

The Partitioning Between Oil and Water

The separation of molecules between oil and water is generally never absolute. By this we mean to say that a particular kind of molecule will dissolve predominantly in one of the phases (oil or water) and a little bit in the other phase. The degree of partitioning varies from molecule to molecule. It turns out that for a given molecule (such as betanine, sugar, or caffeine), the concentrations in the oil phase and water phase have a fixed ratio. For example, if the water phase has X mg/L of betanine, then the oil phase has $0.11X$ mg/L of betanine. Here, 0.11 is the oil–water partition coefficient, K_{ow}, of betanine.

If the partition coefficient K_{ow} has a value larger than 1, then the molecule will have a tendency to end up in the oil phase; such molecules are then also known as lipophilic (fat loving). If K_{ow} is smaller than 1 for a specific molecule, water is the favorite place for this molecule. These molecules are called hydrophilic (water loving). Partition coefficients can be used quantitatively: the lower the value of K_{ow}, the stronger the attraction to water; the higher its value, the more the molecule will accumulate in the oil phase.

Values of the oil–water partition coefficient can be determined experimentally or can be calculated. For the scientifically inclined cook, it might be of interest to know that there are freeware tools on the Internet that estimate the K_{ow} for any compound.

Influence of the Oil:Water Proportions

Actually, multiplying flavor by a factor of 2 by means of a jar of water and oil is a gross *understatement*. In principle, the number of flavors increases by a much larger factor than 2 if one takes into account that each compound has a different oil–water partition coefficient K_{ow}. Every ingredient has a multitude of different kinds of flavor molecules. For instance, coffee has more than eight hundred

Table 1 Concentration and Partition Coefficient of Various Key Components in Freshly Brewed Coffee

Compound	Flavor group	Concentration (mg/L)	K_{ow}
Methylpropanal	Sweet/caramel	28.2	5
2-ethyl-3.5-dimethyl pyrazine	Earthy	0.326	117
2-furfurylthiol	Sulfurous	1.7	91
Guaiacol	Smokey	3.2	21
Fenugreek lactone	Spicy	1.58	0.4
Caffeine	Stimulant	4,800	0.9
Chlorogenic acid	Acids	14,800	0.1
Furfuryl alcohol	Bitters	300	1.9
Rose oil	Fruity	0.23	16,218
Acetaldehyde	Fruity	130	0.5

Table 2 Comparison of the Two Types of Extraction Methods with Different Amounts of Butter

Compound	Flavor group	Concentration in butter 1 mg/L	Concentration in butter 2 mg/L
Methylpropanal	Sweet/caramel	175	40
2-ethyl-3.5-dimethyl pyrazine	Earthy	3.2	2.5
2-furfurylthiol	Sulfurous	16	12
Guiacol	Smokey	28	12.3
Fenugreek lactone	Spicy	1.6	0.2
Caffeine	Stimulant	9,755	1,195
Chlorogenic acid	Acids	4,215	433
Furfuryl alcohol	Bitters	1,090	162
Rose oil	Fruity	2.3	2.3
Acetaldehyde	Fruity	157	18

Note: In case 1, about ½ pound (225 g) of butter was employed directly to extract from 2.2 pounds (1 L) of coffee, whereas case 2 corresponds to about 1 ounce (25 g) of butter to 2.2 pounds (1 L) of coffee for extraction mixed with nonextracted butter to about ½ pound (225 g). This table is by no means exhaustive because coffee contains more than eight hundred flavor compounds.

Sources: H.-D. Belitz, W. Grosch, and P. Schieberle, *Food Chemistry*, trans. M. M. Burghagen, 3rd ed. (Berlin: Springer, 2004); www.coffeeresearch.org.

aroma compounds. Each molecule has a different K_{ow}, and so the partitioning between water and oil is different for each molecule. In table 1, a few important flavor compounds in coffee are given with their subsequent K_{ow}. An interesting consequence of the different values of the partition coefficients of the various flavors is that the relative proportions of the molecules depend on the ratio of water and oil volume that is used in the extraction. This is illustrated in table 2, in which the expected concentrations resulting from two different butterfat extractions of freshly brewed coffee are given.

In figure 46, the effect of phase partitioning is depicted. It shows the dependence of the efficiency of extraction on the oil–water volume for different types of molecules, each having a different partition coefficient. All concentrations are so-called normalized concentrations: the actual concentration is divided by the maximum possible concentration given. This quantity is maximally

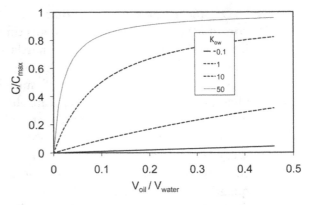

Figure 46 The extraction efficiency (C/C_{max}) is plotted as a function of the ratio of oil to water volume (V_{oil}/V_{water}) for different values of the partition coefficient (K_{ow}).

equal to 1, and lower values indicate less efficient extraction of the compound. It is seen that lower values of the partition coefficient lead to lower efficiency of extraction. Perhaps more important, the less oil is used compared with water, the less of each of the compounds is extracted.

The Role of Temperature

Higher temperatures result in lower viscosity. In particular, the oil phase will generally flow much more easily at a higher temperature. Therefore, the extraction process will become more efficient because the rate of transfer between the two phases will be become faster. Also, some flavors may be trapped in their matrices (for instance, lobster shells). Higher temperatures in combination with water will promote the release of flavors by the weakening of these matrices.

Next to the rate of the extraction, one may also expect that the partitioning between the phases will be affected. First of all, some of the components—in particular, some of the aroma compounds—are not heat stable. Therefore, at higher temperatures, the flavor of an extract may change or be lost, as some of the compounds will degrade as others remain. Therefore, if you want your extract to be as close as possible to the basic ingredient, it is advisable to avoid high temperatures. Another effect is that the partition coefficient may depend on temperature. This will influence not only the extraction rate but also the final partitioning between the phases.

Of practical importance is that when butter is used to extract flavors, it is best to clarify it first. Neglecting to do so can lead to emulsification of the butterfat, making it hard to separate the infused fat. The emulsification is promoted by surface-active proteins that are naturally present in the butter and that will stabilize fat droplets in the water phase, making it difficult to separate the two phases. By clarification, one removes these, and the butterfat will separate much more easily.

Suppose you are not happy with your infused oil or butter because it carries some bitter flavors or some unattractive aromas. One remedy is to wash it out again. This is done by melting the fat in a large quantity of hot water. Then stir vigorously so that the oil droplets finely disperse in the water, allowing the fat to cream again. This way, some compounds that are only somewhat soluble in water (mostly the bitter ones) will escape from the infusion into the water. This is also an excellent way to get rid of some possible particulate impurities in the oil. Note, however, that some interesting flavor is lost in the wash.

The Case of Coffee Extracts

Coffee as a drink contains bitter compounds, acids, and a range of aromatic compounds. In the preparation of a coffee-flavored sauce or ice cream, separation of the aromatics from the bitter compounds is desirable. Based on the oil–water partition coefficients, a good way to proceed is flavor extraction by fat. The bitter compounds will remain in the water phase, whereas most of the nice coffee aromas will be trapped in the fat. As discussed, clarified butter should be used to avoid emulsification.

We have examined two extraction methods. First, we checked what would happen if ground coffee was brought in direct contact with butter. In the second method, the coffee was brewed and then mixed vigorously with clarified butter. In our experience, this method gave the best aromatized butter. In fact, the extract prepared following the first method had a truly bad flavor, which resembled the smell of spent coffee grounds. There are off-flavor compounds that are poorly soluble in water, and these usually remain in the filtrate during the first hot-water extraction. When these compounds come into direct contact with fat, they are immediately absorbed by the fat and thereby render the fat phase highly disagreeable.

The results of a coffee-infused butter prepared using the second extraction method were extremely successful. A very rich coffee aroma was absorbed by the butter with very limited bitter notes taken up by the butter. The watery residue, not surprisingly, was just too disgusting for words!

Next, we put the theory to the test by checking whether the water:fat ratio would influence the quality of the extraction. For this, we made two 2¼-pound (1 L) pots of freshly brewed coffee. To one, we added approximately 1 ounce (25 g) of clarified butter, to the other about ½ pound (250 g). The molten butter was poured into the coffee and stirred for 10 minutes at 176°F (80°C). After preparing the coffee–butter mixtures, we put samples in the fridge to let the oil and water separate and the butter solidify. Next, the butter was scooped off and the fat from the sample initially prepared with about 1 ounce (25 g) of clarified butter was melted with ½ pound (225 g) of non-infused clarified butter. The fat from the sample prepared with approximately ½ pound (225 g) of butter was simply melted. Then, we compared the flavor of the two different samples. In table 2, we tabulated the expected concentrations of a few important coffee-aroma components for the two extraction scenarios.

Our tasting panel consisted of multiple-champion Dutch barista Sander Schat, chocolatier Kees Raat, and beverage experts Johan Kersten and Ruth Lim. We asked them to score for aroma and bitterness. After averaging, it remains statistically inconclusive which extraction was most bitter. On the aroma side, although the extraction with 1 ounce (25 g) of butter was weaker, as suggested by the computations represented in table 1, no differences in aroma bouquet were noted.

To test the influence of temperature, we compared an oil infusion at 176°F (80°C) and one at 86°F (30°C). Again, two 2¼-pound (1 L) pots of

COFFEE BUTTER

17 ounces (0.5 L) very strong freshly brewed hot espresso coffee*
5 ounces (150 g) clarified butter

Put the butter and the coffee together in saucepan and heat to just below the boiling point for 10 minutes. Vigorously blend the butter and coffee with a hand mixer. Heat again and pour the mixture in a bowl. Cool down the mixture to room temperature and put it in the fridge. After a few hours, scoop off the top layer of solid infused butter.

* If you don't have an espresso machine, just prepare 17 ounces (0.5 L) of very strong coffee.

COFFEE SAUCE

7 ounces (200 mL) chicken stock
5 ounces (150 g) coffee butter

Heat chicken stock in a saucepan to 176°F (80°C). Whisk the coffee butter piece by piece into the hot stock. Adjust flavor with salt.

COFFEE ICE CREAM

17 ounces (0.5 L) milk
7 ounces (200 g) sugar
2 g salt
5 egg yolks
5 ounces (150 g) coffee butter, molten

Slowly dissolve the sugar and salt in the milk over low heat to about 170°F (75°C). Put yolks in a stand mixer and emulsify the molten coffee butter into the egg-yolk base. Make sure that the temperature of the molten butter is not above 140°F (60°C) to avoid cooking the yolks. Continuously stirring, add the hot milk to the emulsion. Let the mixture cool to room temperature and continue cooling in the fridge. Pour mixture into the ice cream maker.

coffee were brewed. Each sample was subsequently mixed with 3½ ounces (100 mL) of sunflower oil. Afterward, one sample was mixed for 10 minutes at 86°F (30°C) and the other one for the same amount of time, but at 176°F (80°C). The oil in both samples was creamed and cooled. We then compared them. In this comparison, the panelists were inconclusive on bitterness, but the aroma quality of the hot extraction was milder and fruitier than the cold extraction, which was distinctly heavier.

We also analyzed the caffeine content of the butter and found that, as expected, it was less than the caffeine found in straight coffee. This was consistent with the oil-to-water partition coefficient of caffeine; that is, it shows a preference for the water phase.

In conclusion, we praise infused oils and butters for what they lend to cooking and we provide two recipes using coffee-flavored butter. The method of extraction for general kitchen use will have an influence on the flavor. It is difficult to say in general what the best way is to optimally invoke the technique of solvent extraction. But we found that the highest yields of flavor are obtained by using relatively large amounts of fat or oil. We therefore advise using exactly the amount of fat the recipe requires. It is the most economical way to proceed.

As you can see, molecular gastronomy has truly inspired us. It starts with an idea. In this case, we strove to separate flavors by simultaneous water–oil extraction. And this inspires the research—to postulate ideas and test them in a systematic kitchen setting. The proof of the pudding is always in the eating. Therefore, tastings by aficionados is the way to ultimately decide whether it makes sense to actually use a particular idea. Ultimately, we just hope that you will enjoy the accompanying recipes and that we have inspired you to apply these concepts in the extraction of other flavors.

Further Reading

"Coffee." 2010. http://www.coffeeresearch.org.

Gagnaire, Pierre, and Hervé This. 2005. "L'enfleurage des fromages." Art et Science. Le travail du mois. http://www.pierre-gagnaire.com/francais/cdthis.htm.

Mariën, Eke, and Jan Groenewold. 2007. *Cook & Chemist*. Edited by Bas Husslage. Uithoorn: Karakter.

——. 2008. *Meer recepten uit de moleculaire keuken van Cook & Chemist*. Uithoorn: Karaker.

Peterson, James. 2008. *Sauces: Classical and Contemporary Sauce Making*. 3rd ed. Hoboken, N.J.: Wiley.

Pybus, David H., and Charles S. Sell, comps. 2004. *The Chemistry of Fragrances*. Cambridge: Royal Society of Chemistry.

Reineccius, Gary A. 2006. *Flavor Chemistry and Technology*. 2nd ed. Boca Raton, Fla.: CRC Press.

This, Hervé. 1995a. "La gastronomie moléculaire." *L'Actualité chimique* 5–6:42–46.

——. 1995b. *Révélations gastronomiques*. Paris: Belin.

twenty-six

TURNING WASTE INTO WEALTH

On Bones, Stocks, and Sauce Reductions

JOB UBBINK

one of the things that immediately captivated me when, as a student, I started working part time in the kitchen of a French restaurant was the hectic pace in which a ten-person kitchen brigade worked to get the food out on time during the evening service. The restaurant was located in a small village near the university town of Leiden, where I was pursuing my studies in physical chemistry. At that time—the end of the 1980s—interest in food, cooking, and gastronomy was rising very rapidly in the Netherlands, a country that was still commonly seen as a gastronomic backwater. The restaurant was doing very brisk business, particularly on Thursday, Friday, and Saturday evenings—my regular shifts.

Toward the end of the evening, with only dessert left to serve, a more calm and relaxed atmosphere would settle into the kitchen. Rather suddenly, the pressure under which we had been working would drop, and attention would gradually shift to the preparations for the next day, in particular, to the making of stock. Kitchen assistants—like me—would take bones out of the walk-in refrigerator and clean onions, carrots, and celery roots. We would heat up the oven to quickly roast the bones and vegetables on baking sheets, before bringing them to a boil, along with spices and water, in huge stainless steel pots. As they simmered overnight, the veal, poultry, or beef stocks would release their complex, fragrant aromas throughout the kitchen. In the morning (when I was in the chemistry lab—not even half as interesting!), the stocks would be filtered through cheesecloth, cooled down, and defatted. Then—pure magic—they were turned into the incredibly rich, viscous, golden-brown stock reductions known as glaces or demi-glaces by vigorous boiling on high heat to evaporate the water and concentrate their flavors (figure 47).

Why did the cooking of stocks and reductions fascinate me so much? Was it largely the smell, color, and texture of the product? Or was it also the feeling that this was true cooking, with respect for the ingredients and rooted in a venerable tradition? It was not something I knew from eating at home as a kid; I was also not specifically attracted to the richness radiating from the dishes the sauces were accompanying. I was simply fascinated by the process of extracting something this rich, delicate, and tasty from products that, before my foray in a restaurant kitchen, I considered to be little more than waste—bones and tough connective tissue.

Inspired, I started experimenting with stock making and the reduction of stock to glaces and demi-glaces in the kitchen of my dormitory—this to the amazement of my fellow students, who, in the same kitchen, would usually prepare nothing more elaborate than spaghetti or lasagna. When creating sauce reductions, I would follow my purist instincts and use only those ingredients I deemed to be "pure": the flavor and taste but also the color and texture of the sauce should be derived only from the original stock, and this in a fine balance. How great was my disappointment when, one day at lunch service in the restaurant, I saw the sous-chef, just before service, add a small teaspoon of a dilute starch dispersion to a demi-glace, just to "top up the texture." To me, this was treason, foul play, very much unethical cooking—and completely unnecessary.

Since those days, I have learned a lot about the basic science of food and food preparation. As a consequence, I have become more open-minded about how to cook, including how to prepare stocks and their derived sauces. I am still sticking to my old beliefs in the use of "pure" ingredients, but I have started experimenting with how to perform the extraction. In so doing, I have begun to play with the balance between the reduction of the volume of the stock, the release of the volatile aroma compounds, and the generation by thermal reaction of new, additional flavors and taste compounds.

Figure 47 Classical preparation of veal stock and demi-glace: (*top*) roasted bones and vegetables; ingredients together in the pot; (*center*) stock after simmering and then after defatting and filtration; (*bottom*) stock reducing and then the final product: demi-glace.

What, then, are the key physical and chemical steps in the preparation of stock and sauce reductions? Surprisingly, fairly little is known about this. Of course, people such as Justus von Liebig long ago investigated how meat extracts can be prepared most economically and much is known about the extraction of gelatin and the properties of the extracted gelatin. But the processes that occur in stock making and reduction have never been investigated in detail.[1]

The process of stock making and reduction to a demi-glace consists of three steps. We start by extracting the ingredients (bones, meat, connective tissue, vegetables, herbs and spices). Filtration and separation follow, during which the remaining solid ingredients as well as the extracted fat are removed. In the third step, the stock is "reduced" to the concentrated sauce. Scientifically, various physical and chemical phenomena play important roles in the process of stock preparation and reduction: extraction, hydrolysis, chemical reactions, phase separation, and phase partitioning.

The first step, extraction, consists of not only the liberation and release of compounds already present in the connective tissue but the hydrolysis and breakdown of collagen in particular. Collagen is the main protein of connective tissue, such as cartilage, ligaments, and tendons. As in all proteins, collagen is made of a linear chain of amino acids (a so-called polypeptide). In the collagen fibers, these polypeptide chains are, to a variable degree, chemically linked together, forming a cross-linked network that confers toughness and resilience to the fibers.

Apart from gelatin, we extract other components from the connective tissue. What precisely these other components are depends on how much muscle fiber the base material still contains, but the extract will include free amino acids, ribonucleotides, phospholipids, minerals, and some sugars. The sugars derive from glycogen, which is a polymerized version of glucose and the form in which animals store carbohydrates. Sugars are also extracted from the vegetables added to the stock: sucrose from the carrots and fructose from the inulin present as the storage carbohydrate in onions (the sweet taste that onions develop when slowly cooked is from the fructose that results from the depolymerization of polyfructose, also known as inulin).

When making stock, we need to know how fast the various components are extracted. If the extraction occurs too rapidly, we risk leaving behind a significant fraction of the interesting ingredients in the bones and vegetables at the end of the process. Moreover, we do not want to lose too many of the interesting flavor and taste compounds that evaporate from the stock or that degrade from the chemical reactions that occur during the extraction. Without these

1 The scientific literature is very limited; some of the principal references can be found at the end of the chapter.

concentrated flavors, our sauce would be rather bland. Thus, how long should we let our bouillon simmer? Chefs generally have divergent opinions. Often, they say that extractions should be quite short or the sauce loses its quality and becomes too heavy (see Michel Roux's [1996] wonderful book on sauces, for example). However, the longer we extract, the more extensive the hydrolysis and extraction of the collagen, which provide both the gelatin polymers needed to give body to the sauce and the short peptides that can participate in flavor-generating reactions.

The duration of extraction depends on the types of bones and the varying sizes of the ingredients we are using: fish bones and waste will generally require shorter extraction times than veal because of the looser structure of their collagen and their smaller size. Conversely, because collagen from beef tends to be of higher molecular weight and more intensively linked together, beef stock will require a longer extraction time than veal stock.

We may safely anticipate that the hydrolysis and solubilization of the collagen is the slowest step in the making of bouillon, as it involves the breakdown of a chemically quite stable material, whereas the other compounds we extract into the stock are simple low-molecular-weight compounds locked up in the interior of the muscle cells. Consequently, there is no need to continue the extraction for longer than is needed to effectively extract the gelatin. This aspect of stock making has luckily been studied in the context of industrial gelatin manufacturing: a good balance between the amount of gelatin extracted and the molecular weight of the gelatin is reached under conditions that, translated to the kitchen, give a maximum extraction time for connective tissue from normal-size beef or veal pieces (typically 1¼ to 2 inches [3–5 cm] diameter) of 5 to 7 hours. General recommendations for the optimal extraction time are very rough; it will depend on many other conditions—in particular, the acidity or pH of the stock. But overall, it gives us a reasonable figure to work with, and we can conclude that overnight extractions, of, for instance, 12 to 14 hours, do not add body to the stock and could result in the loss of valuable compounds. Very long extraction times will lead to a reduced body for the stock, as the gelatin will further break down, giving rise to a decreasing viscosity, even though the gelatin concentration is still increasing. Harold McGee (2004) confirms that whereas after 8 hours of extraction (of beef), only about 20 percent of the gelatin is extracted, the body of the stock decreases for longer extraction times.

The maximum extraction time of 5 to 7 hours for veal or beef can be influenced by changing the extraction conditions: if we were to use a pressure cooker to extract the bones, the increased cooking pressure (about 2 bars) would result in higher cooking temperatures (between 250 and 265°F [120–130°C]), significantly shortening the extraction time. Using as a rule of thumb that chemical reactions double in speed for every 18°F (10°C) increase in temperature, optimum extraction times will be about 1 to 2 hours. We could also presume that roasting the bones prior to the extraction will significantly influence the extraction time. I do not believe this to be true, however, because during short roasting times (for stocks, the ingredients are typically roasted 10 to 20 minutes at 390 to 430°F [200–220°C]), the collagen does not significantly degrade—within the bones, the temperature hardly rises during the short roasting time.

Is a period of 5 to 7 hours perhaps already too long for an optimal extraction? It might be, and as many chefs know, we lose interesting compounds from our stocks at longer extraction times. Therefore, to have a good extraction of at least part of the gelatin, along with interesting flavor notes, we could perform a double extraction, using a new batch of ingredients for a second, shorter extraction.

So, how much can we extract from the bones? This, again, will vary with the type of animal tissue used and depend on whether it contains a significant amount of bones. Bones, being a highly mineralized tissue, do not appreciably degrade during the preparation of the stock and will only release a limited amount of gelatin. The bone marrow is different and will release compounds, but only when the bone is cut or fractured to free it up into the stock. In general, most of the compounds of importance for our stock are extracted from the connective tissue and the meat muscles.

One compound that normally is much appreciated for the texture, flavor, and mouthfeel it brings to foods, but is a hassle in stock, is fat. It easily melts and is easily extracted into the stock. Interestingly, whereas many flavor and aroma compounds usually like to accumulate in a fat phase (a phenomenon that is called phase partitioning, as discussed in chapter 25), we do not observe this for many of the compounds in stock. Refrigerate a freshly extracted beef stock for a couple of hours, then taste some of the greasy, dirty, yellow-whitish layer that accumulates and solidifies on top: it will not taste very good, and it will contain virtually no sign of the delicate and complex taste and smell associated with the stock. Many of the important taste and aroma compounds are thus mainly water soluble. And the fat? Even though it is easy to remove from the cooled-down stock because it nicely phase-separates and its lower density allows it to settle at the top, we should try to remove as much as possible from our starting material. This is because it may oxidize during the extraction and form unpleasant flavor notes that will linger in the final product.

In my experience, the amount of nonfat dry matter solubilized in stock is between 2 and 5 percent by weight, depending on the starting ratio between bones, skin, and meat on the one hand and water on the other (I often use a ratio of 1:2). Assuming that somewhat less than half of the weight of the bones consists of collagen-rich connective tissue, about 10 to 20 percent of the connective tissue is extracted during stock making. Of all extracted matter, only a very small fraction (much less than 1 percent) consists of the volatile aroma compounds we perceive when a stock is gently heated on the stove.

Are there innovative ways to change or even improve the preparation of stock? I already mentioned the effect of pH: often during extraction, the pH is around 4 to 5; that is, the stock is slightly acidic (wine or a tomato has been added to it, for example). We could prepare our stock as a base, which would be at a pH of more than 7. To my knowledge, what happens at this pH is not known; I will try a higher pH in the future! Alternatively, we could pretreat the meat with an enzyme, for instance a protease or collagenase, which hydrolyzes the proteins in the meat tissue before extraction and speeds up the extraction process. In so doing, however, we are likely to end up with an unfavorable stock composition that could have a very strong negative influence on the taste and smell of the reduction.

We now have our stock. It is properly extracted, filtered, and defatted, and we want to turn it into a glace or demi-glace to serve as the basis for a fine French sauce. For this purpose, a chef, as I mentioned, boils down the stock to the desired consistency, color, and taste. For a demi-glace, the volume of stock is typically reduced by a factor of two; for a glace, the volume reduction can be anywhere between 70 and 90 percent. It is largely this process, which in culinary terms is called the "reduction," that captured my fascination because of the delightful smell it produces. However, this delightful smell also means that many attractive volatiles are lost.

Can we improve on this? What about using a concentration step that avoids heating up the stock and losing these wonderful volatiles? Highly effective are the low-temperature concentration processes that the food industry has been using for a long time, such as freeze drying, osmotic concentration, and freeze concentration.

Figure 48 Freeze drying of stock (a recipe modification): (*from left to right*) stock in the freeze dryer; dehydrated stock; milling of the freeze-dried stock; reconstituted freeze-dried demi-glace.

Over the past few years, I have been experimenting with freeze drying (which consists of first freezing the product and then drying it in a vacuum) in the preparation of stock reductions with good results (figure 48). Upon freezing a stock, ice will form at temperatures below the freezing point—slightly less than 32°F (0°C), depending on the composition and the concentration of the stock (figure 49). The ice is a pure phase consisting only of water; upon the formation of ice crystals in the stock, the remaining, nonfrozen part of the stock will become much more concentrated. This process, which is known as cryoconcentration, leaves us with a concentrated stock with dispersed ice crystals. A similar but natural process to concentrate flavors and sugars is traditionally used for ice wine, in which the grapes are left to freeze on the vine. We can go even further by drying the frozen extract and sublimating the ice crystals from the concentrated stock, leaving a highly concentrated stock phase. This sublimation of

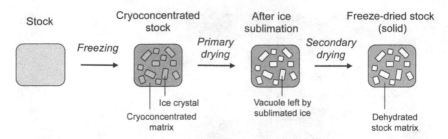

Stock   Cryoconcentrated   After ice   Freeze-dried stock
             stock   sublimation   (solid)

Freezing   *Primary*   *Secondary*
                       drying   *drying*

Ice crystal   Vacuole left by
Cryoconcentrated   sublimated ice   Dehydrated
matrix   stock matrix

Figure 49 The steps in the reduction of stock by freeze drying.

the ice crystals during freeze drying is done under very low pressure (typically, a few millibars) to speed up the process and is known as the primary drying.

Because we are not looking for a completely dry product, we could in principle stop the freeze-drying process when most of the ice crystals are sublimated, which would result in a very concentrated, pastelike sauce reduction. However, I typically continue with what is called the secondary-drying phase. In this phase, most of the water in the cryoconcentrated-sauce phase is also evaporated. If we continue the secondary drying for long enough (typically 1 to 2 days), we can end up with a solid, porous layer of sauce, resembling—to some extent—the structure of materials used in home insulation (top-right photo in figure 48). This "solid sauce" is fragile and can easily be milled into a powder (bottom-left photo in figure 48), which in turn can be dissolved in water to obtain a reduction: a glace or demi-glace, depending on the water content, with a body and mouthfeel that are very similar to those of sauces prepared by traditional methods (bottom-right photo in figure 48). In principle, the "solid sauce" can also be used as it is because it easily melts in the mouth, giving a nice release of flavor.

The freeze-dried, reconstituted sauce is quite different from the equivalent sauce prepared by reduction over heat. It contains much more of the aroma and taste notes associated with the original ingredients. The glace is thus more meaty, and more of the vegetable and spice notes are apparent, while the strong caramelized flavor and smell usually associated with thermally reduced sauces have decreased or even disappeared. Overall, the freeze-dried reductions are lighter and more delicate than the classical sauces, but some of the typical meaty and nutty flavor notes are lacking. This is because during the reduction step, reactions occur, for instance between the reducing sugars and the amino acids. These reactions are collectively called the Maillard reaction; it gives rise to flavors, volatiles, and colored compounds (as discussed in detail in chapter 13). Therefore, we do not only extract aroma and taste compounds during the extraction and reduction of the stock, but we also generate them.

In the preparation of the optimal glace or demi-glace, we thus need both a concentration step and a reaction step, both traditionally achieved through thermal reduction. By using freeze drying followed by a short heating up of the reconstituted sauce (with a lid on the saucepan to keep the delicate volatiles from escaping), we effectively decouple the concentration, or reduction, step and the reaction step, and we optimize both steps individually.

Are there other ways in which we could play with the reactions in the sauce and thus further modulate its taste and flavor profile? Here I am hesitant. As I mentioned, for some of the stocks, the bones and vegetables are roasted to release flavor compounds before the extraction. We can add vegetables containing different sugars to our stock to change the type and content of what are called the precursors of the Maillard reaction and, consequently, the volatile profile of the stocks and sauces. These precursors can also be added as separate components, in the form of pure sugars and amino acids, as is routinely done in the food industry. By adding such pure compounds, we can much better control individual steps in the creation of the flavor. However, we also risk losing the delicate balance between the many flavor compounds that make up the overall aroma and taste of the

Figure 50 The final meal: filet et cuisse de pigeon sauté avec sa sauce.

sauces. Therefore, I am not in favor of using such simple building blocks to compose the flavor profile of a complex stock or sauce.

So where does this leave me? Starting out as a young adult convinced that cooking was an almost sacrosanct endeavor bound by recipes and manipulations fixed in time-honored traditions, I have become more aware of how significant change has been in our appreciation of foods and dishes. However, I am still very much convinced that ingredients have intrinsic qualities that we should respect and that any modern food technology should not be used to cut corners in the preparation of cheap, unhealthy fast food but rather to better bring out the intrinsic qualities of the "real" food—maintaining a vibrant, sustainable food culture. And I am still preparing my traditional sauces (figure 50), only now, occasionally, with a modern technological twist.

Further Reading

Cambrero, M. I., I. Seuss, and K. O. Honikel. 1992. "Flavor Compounds of Beef Broth as Affected by Cooking Temperature." *Journal of Food Science* 57:1285–1290.

Guth, H., and W. Grosch. 1994. "Identification of the Character Impact Odorants of Stewed Beef Juice by Instrumental Analyses and Sensory Studies." *Journal of Agricultural and Food Chemistry* 42:2862–2866.

McGee, Harold. 2004. *On Food and Cooking: The Science and Lore of the Kitchen*. 2nd ed. New York: Scribner.

Petersen, James. 2008. *Sauces: Classical and Contemporary Sauce Making*. 3rd ed. Hoboken, N.J.: Wiley.

Roux, Michel. 1996. *Sauces: Sweet and Savory, Classic and New*. London: Quadrille.

Snitkjær, P., M. B. Frøst, L. H. Skibsted, and J. Risbo. 2010. "Flavour Development During Beef Stock Reduction." *Food Chemistry* 122:645–655.

twenty-seven

RESTRUCTURING PIG TROTTERS

Fine Chemistry Supporting the Creative Culinary Process

JORGE RUIZ AND JULIA CALVARRO

nose-to-tail eating is a sort of culinary philosophy based on the premise that nearly every part of the animals we sacrifice, in the hands of patient and talented cooks, can be made into a rewarding eating experience. Nose-to-tail eating is nothing new, however, as it is common around the globe. For instance, Mediterraneans, such as the French, Italians, and Spanish, as well as the Chinese and Mexicans are known for eating the ears, hearts, livers, cheeks, tails, brains, and feet of pigs, cows, lambs, and other animals.

The Recipe and the Problem

This story is about pig trotters—*stuffed* pig trotters.

Boiled pig trotters (pig's feet) with tomato, garlic, onion, bay leaves, parsley, pepper, blood, and fermented sausage (chorizo) is a traditional dish in Extremadura and several other regions of Spain. A modified version of this delicious dish is part of the menu served at Restaurant Atrio, in Cáceres, Spain, whose kitchen is run by the talented chef Antonio "Toño" Perez. His version of this dish is as challenging as it is scrumptious. The challenge revolves around the lack of structure of the cooked trotters: they frequently fall apart during cooking, which makes them unsuitable for serving. We visited

Perez to discuss the details of his culinary problem with the intent of finding a solution. It was obvious that the disintegration of the trotter during cooking could not be avoided. In fact, the final texture of the dish relies on a thorough softening of the trotter. This led us to suggest a series of restructuring strategies that, within the context of this dish, could be referred to as meat glues.

A long cooking process, usually about 2 hours, is necessary to soften pig trotters. With their many tendons and ligaments and thick skin, trotters have ample amounts of collagen that must be broken down. During this prolonged cooking time, collagen is transformed into gelatin, which gives the pig's feet their peculiar soft texture when warm and their gumminess and stickiness when cold (for more details on the properties of collagen, see chapter 23).

As you can imagine, the textural features of pig trotters are fascinating from a chef's point of view. This prompted Perez to take a different approach to the traditional recipe. After cooking them for 2 hours, and while they are still hot, he butterflies the trotters (that is, opens them longitudinally, skin side down) and removes the small bones and cartilage (figure 51a). He then lays an open trotter on top of a square piece of plastic wrap and places a portion of raw foie gras and a couple of pieces of sautéed ceps (mushrooms) in the middle of the trotter. A couple of tablespoons of pork stock are poured onto the trotter before rolling and wrapping it in the plastic. The "trotter roll" is then wrapped two or three more times to ensure a fairly compacted structure.

The roll is then refrigerated, and as it cools down, the gelatin acts as a glue that holds the meat cylinder together. When an order for the dish is called for, the cooks cut a medallion about ¾ inch (2 cm) thick, remove the surrounding plastic, and grill it until a nice crust forms on both sides. Aromas of delicious roasted pig skin fill the kitchen (figure 51d). To serve, the chef pours a small amount of a reduced stock onto the plate, places the medallion on top, and garnishes with some green vegetables and a small piece of the outer layers of a leek (figure 51e). The dish possesses the particular soft texture of the trotters, the roasted flavor developed during searing, the richness of the concentrated stock, and the lightness of the vegetables—it is delicious!

You may wonder, where is the challenge we talked about? The issue was that the shape of the medallion was lost during or after grilling, a problem that seemed to worsen when there was not an even layer of skin around the medallion. Let us think about what happens. During grilling, two competing and antagonist phenomena occur: on the one hand, the crust that forms at the surface eventu-

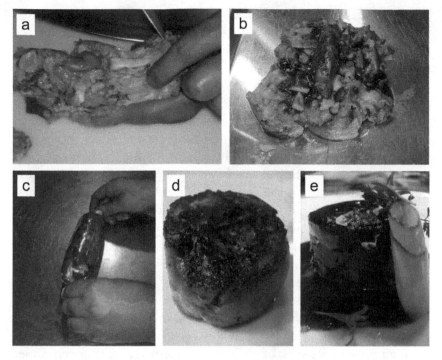

Figure 51 Steps in the elaboration of stuffed trotters: (*a*) deboning and removing of cartilage from pig trotters once they have been boiled in water for nearly 2 hours; (*b*) view of the boiled and deboned trotters, opened and with the foie gras and mushrooms on top—and the meat stock poured over; (c) wrapping the trotters with the plastic film; (*d*) grilled trotters; (*e*) finished and plated dish.

ally helps hold the medallion together; on the other, as the interior warms up, the gelatin—the glue that maintains the medallion's structure—begins to melt. If the gelatin melts before a strong enough crust forms, the medallion loses its shape. This was a common occurrence. Needless to say, deformed medallions meant significant waste and stress during the preparation of the dish.

Potential Strategies for Restructuring

Considering that the cooked trotters possess a high level of protein in the form of gelatin, we hypothesized that the use of "meat glue" could represent an alternative means to keep the shape of the medallions during cooking.

Microbial transglutaminase (TGase), known in the culinary world as "meat glue," is an enzyme that induces the formation of stable bonds between protein molecules. This property is indeed exploited by the food industry: TGase is used to bind pieces of meat for the manufacture of, for example, deli meats. Gelatin is actually highly reactive to the action of TGase.

Another very common way to restructure fish or meat is based on the formation of a calcium alginate network. This procedure relies on the ability of sodium alginate to form thermally irreversible calcium alginate networks (for more details about the chemistry of alginate gels, see chapter 4). However, in the case of binding systems for meat or fish, an internal setting or in situ gelation occurs. To visualize how this works, think about the magic of dental impressions. The dental technician combines a mysterious powder with water to form a paste. In as little as 2 to 3 minutes, this paste sets to a firm solid, revealing every detail of our teeth. This is a sophisticated application of internal setting gels, which rely on a fine balance between pH change and calcium (Ca^{2+}) and alginate solubility.

A third restructuring strategy is the use of fibrinogen–thrombin networks, which are the proteins involved in the formation of blood clots. The binding mechanism in this case is an insoluble and heat-resistant protein network of fibrin that is formed as fibrinogen comes into contact with thrombin. There are commercial kits available that offer either concentrated solutions or freeze-dried powders of fibrinogen and thrombin—typically isolated from the blood of cattle and pigs slaughtered at highly regulated facilities. We chose to use a powdered plasma preparation (Fibrimex [FIB]) containing premixed fibrinogen and thrombin. The reaction takes place as soon as the mix is dissolved in water.

Restructuring Step by Step

Transglutaminase

We prepared a solution of about ½ ounce (12.5 g) of TGase (Ajinomoto, Japan) in 1¾ ounces (50 mL) of the water in which the trotters were cooked. We used this water because, being rich in gelatin, it is a great substrate for the enzyme. Once the trotters were stuffed with foie gras and mushrooms, we poured approximately 1½ ounces (45 mL) of the TGase solution over the trotters, trying to cover as much surface area as possible. We then proceeded to wrap the trotters as described earlier.

Fibrimex Plasma Powder

In using the plasma powder, we followed the exact same procedure as for TGase, but in this case, we used ½ ounce (12.5 g) of plasma powder.

Alginate/Calcium Sulfate (ALG)

We used a two-envelope commercial powder preparation, one envelope containing sodium alginate and the other calcium sulfate. We sprinkled sodium alginate at about 1.2 percent of the trotter's weight onto the surface of the stuffed trotters. We prepared a 2 percent solution of calcium sulfate and then poured 1½ ounces (45 mL) of it on top of the trotters as described earlier.

The Results

The chef cooked the trotters made by these three restructuring methods. A control batch of trotters, made without a binding agent, was also prepared as a way to compare outcomes during the grilling step. We tasted the result without any garnish or sauce so as to fully experience the texture profiles of the restructured meat and to avoid masking any potential off-flavors derived from the glues used.

The ALG and FIB systems resulted in medallions with much better cohesiveness than those restructured with TGase and without any glue (the control). Figure 52 shows a trotter restructured with ALG before, during, and after grilling. You can observe that the medallion has a piece of mushroom on its surface. In the absence of glue, this medallion would have probably fallen apart; however, even in this worst-case scenario (that is, a piece of mushroom at the surface), the restructuring strategy paid off. Given that the calcium alginate network is thermally stable at very high temperatures, the gel that formed did not melt when heated.

Based on what scientists have previously reported, it was surprising to see the poor results obtained with TGase. The binding effect of TGase is based on the formation of chemical bonds between proteins, and other researchers have proven that gelatin shows a good reactivity with TGase. We are not sure how to explain the poor performance of TGase, but we believe that part of the problem is the fat contained

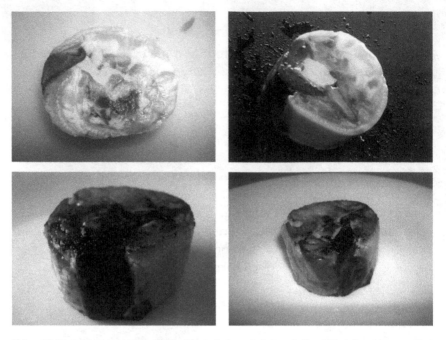

Figure 52 Restructured pig trotters with alginate before, during, and after grilling. It can be seen that each trotter has a piece of mushroom on its surface, and despite this, it keeps its shape after grilling. The mushroom is the darker element visible at the edge of the top surface of the trotters.

in the trotters. The fat might have obstructed the access of TGase to the gelatin. More experiments are necessary.

As for the texture of the final products, no drawback was perceived. This was in line with the results shown by consumer studies comparing the use of FIB and ALG in restructured beef steaks. With regard to flavor, even though previous experience suggested we might find soapy and bitter notes in the ALG medallions, most likely due to the calcium sulfate salt, this was not the case. However, some chefs claimed they could detect it. As for FIB, it is not uncommon to detect some off-flavors in restructured meat products; luckily, given the already strong flavor of the trotters, no undesirable flavors were reported.

Considering all the results, as well as the practical comments from the chefs about the potential use of these ingredients in the kitchen, we selected the ALG system as the best for the restructuring of pig trotters. It provides the necessary cohesiveness for grilling without affecting the overall texture or flavor profile. FIB also resulted in very good cohesiveness, but its use in a kitchen context is not very practical—once the powders are dissolved in the stock, the FIB mixture must be poured immediately onto the trotter, otherwise the fibrin network forms prematurely. Besides, chefs are more open to using ingredients derived from algae (alginates) than from blood (fibrinogen plus thrombin), even though they know we eat the latter as part of one of the most traditional foods in Spain: blood sausage.

A Final Reflection

We have demonstrated how specific chemical reactions are highly useful tools in the restructuring of culinary preparations. This has made us fully aware of the fact that cooking is chemistry. Also, we have provided evidence that the application of new technologies and/ or ingredients to culinary processes is valuable not only for routine problem solving but also for the creation of new and highly innovative recipes. This is yet more proof that we can achieve better cooking through science!

Acknowledgment

This research was partly supported with the project funded by the Junta de Extremadura "Estudio y mejora de procesos culinarios empleados en la alta restauración" (3PR05B032).

Further Reading

Flores, N. C., E. A. E. Boyle, and C. L. Kastner. 2007. "Instrumental and Consumer Evaluation of Pork Restructured with Activa™ or with Fibrimex™ Formulated with and Without Phosphate." *LWT–Food Science and Techology* 40:179–185.

Henderson, Fergus. 2004. *Nose to Tail Eating: A Kind of British Cooking*. London: Bloomsbury.

Kamozawa, Aki, and H. Alexander Talbot. 2010. "Transglutaminase." In *Ideas in Food: Great Recipes and Why They Work*, 278–288. New York: Clarkson Potter.

Suklim, K., G. J. Flick, J. E. Marcy, W. N. Eigel, C. G. Haugh, and L. A. Granata. 2004. "Effect of Cold-Set Binders: Alginates and Microbial Transglutaminase on the Physical Properties of Restructured Scallops." *Journal of Texture Studies* 35:634–642.

twenty-eight

INNOVATE

Old World Pizza Crust with New World Ingredients

THOMAS M. TONGUE JR.

as a third-generation american of Italian heritage, one of my favorite meals was my mom's homemade pizza. The dough was thick, chewy, and topped with my favorite ingredients on one half and my brother's on the other. Mom was in the kitchen all day preparing the dough for the crust. The wonderful aroma of fresh-baked pizza that filled the house quickly brought everyone to the table for dinner.

Today, when I cook my mom's pizza recipe for my family and friends, thanks to new technologies that yield novel ingredients, I can produce her Old World culinary delight with a new twist. I do not have to spend my entire day preparing the dough or accept the variable results caused by time delays when people run late against my well-planned timetable (because yeast-raised dough cannot wait).

There are many stories surrounding the origin of pizza (Greece versus Italy), although I was always told that pizza was developed in the sixteenth century as a white-sauce flatbread dish that was sold in the streets of Naples. By the seventeenth century, it had evolved to the oil-coated flatbread with cheese and tomatoes that is the basis of what we recognize today as pizza. American soldiers returning from Europe after World War II brought home their taste for pizza, and

Table 1 Mom's Pizza Crust Recipe (Midwest Formula)

Ingredient	Amount	Percentage
Bread flour	11 ounces (330 g)	60.4
Warm water (82°F [28°C])	6 ounces (180 g)	34.3
Granulated sugar	10 g	1.8
Vegetable oil (olive, canola)	10 g	1.8
Salt	7 g	1.3
Active dry yeast	2 g	0.4

the first pizzerias in the United States began expanding toward the end of the 1940s. The Americanization of pizza evolved into some wonderful regional specialties, including New York–style thin crust, Chicago deep dish, and California-style, with its toppings that challenge the most creative imaginations. This Western pizza has also taken the form of some not-so-wonderful fast-food products and the fully unappetizing frozen grocery pies. Thankfully, today's degradation of pizza into fast food has been interrupted by the innovation of self-rising and take-and-bake pizza. These innovations combine great crusts, high-quality ingredients, and the benefits of home baking, although they still do not achieve a truly homemade result—the pizza I enjoy making, eating, and serving to my family and friends!

My mom's homemade pizza recipe is relatively simple (table 1), although it is very time consuming to make. Like a quality pizzeria restaurant, she made each dough (1- by 14-inch crust) one at a time, by hand, first thing in the morning. The early start allowed the gluten to form, the yeast to ferment a bit, and the dough time to proof and rest so it would roll out evenly into a thick or thin crust. The downside to this process is that yeast is a living ingredient with a life cycle. Variations in the performance of the leavening system can change the rise thickness and the density and chewiness of the dough, depending on what time you finally eat.

While I always appreciated my mom's recipe, I first needed to determine why she chose each ingredient before considering the alternatives to improve upon it. Let us briefly review the essential ingredients.

Flour is the "meat" of the dough recipe. Chefs use high-gluten flour because gluten makes the crust strong enough to stand up to the water and other ingredients. The main difference among flour types is in the gluten content, which varies depending on whether the flour is made from hard wheat or soft wheat. Gluten is the protein that helps yeast stretch and rise. To achieve the best baking results, use the type of flour that best suits the crust you prefer:

- *Bread flour* is an unbleached, high-gluten blend of mostly hard wheat and is best used in yeast breads, as well as in traditional and thin-crust pizza.
- *All-purpose flour* is designed for a number of uses, including cookies, quick breads, biscuits, and cakes. A mixture of high-gluten hard wheat and low-gluten soft wheat, it comes in both bleached and unbleached forms, which can be used interchangeably. This makes it suitable for deep-dish and cracker-crust pizza.
- *Self-rising flour* is all-purpose flour that has had baking powder and salt added to it. This is best used in quick-bread recipes because it already contains a leavening ingredient (baking powder).
- *Cake flour* is made predominantly of soft wheat. Its fine texture and high starch content make it ideal for use in tender cakes, cookies, biscuits, and pastries that do not need to stretch or rise much.
- *Pastry flour* is similar to cake flour, although it has a slightly higher gluten content. This contributes to its elasticity, which is needed to hold together the buttery layers in flaky baked goods, such as croissants, puff pastry, and pie crusts.

During my testing, I found that bread flour achieved the best crust for my traditional taste (table 2). You, too, should try various flours to achieve a texture that is to your taste. However, trust me, your friends and family will like high-gluten bread flour crusts the best.

The second ingredient in pizza dough is water, which typically makes up 35 to 55 percent of the dough. Water is vital to the recipe because it helps disperse and dissolve the ingredients, facilitating their mixing. It hydrates both the starch and the gluten, which helps the dough develop an elastic network during kneading. Also, water dissolves the sugar, allows the chemical leavening to function, and controls the temperature of the dough. The dough can be made to be pliable or stiff, so I strongly suggest that you experiment with different water percentages to achieve the pliability and texture you prefer.

Sugar is typically added up to a level of 5 percent. It feeds the fermentation process. More sugar equals quicker fermentation (which is not necessarily a good thing). Sugar also helps the crust to brown and keeps the water inside the crust.

Table 2 Flour Types and Some of Their Properties (Relevant to Pizza Making)

Compound	Protein level	Gluten strength	Dough strength	Water-absorbing capacity
All-purpose flour	10–12%	Low	Soft	48–52%
Bread flour	12–13%	Medium	Medium	51–56%
High-gluten flour	13–14%	High	Stiff	55–60%

Vegetable oil or shortening ranges in use up to 5 percent. The fat increases the pliability and softness of the dough by providing lubrication. Some chefs use canola or olive oil to add a little extra flavor, although other choices are functionally effective.

Salt is typically used at levels of between 1 and 3 percent. Salt provides extra flavor and impacts the texture of the dough as it interacts with the gluten proteins. It also slows the fermentation process in yeast-raised dough.

Varying these ingredients is simple and delectable, so be inventive and keep an open mind. Do not hesitate to start with either my mom's proven recipe, or mine (which appears at the end of the chapter). Keep your eyes open and your ears perked to observe the reaction of your family and guests, and continue to make it your own.

Let us now consider alternative leavening options to expand your dough preparation time while maintaining the quality of your crust.

Leavening systems (biological or chemical) provide texture and softness by producing carbon dioxide (CO_2) gas bubbles that enable the dough to rise. If you pick up a slice of pizza and examine the crust closely, you can see that it is full of air holes. The production of CO_2 during the proofing of the dough is largely what is responsible for the texture of your crust. Leavening systems can be biological or chemical, depending on whether the carbon dioxide is produced by living organisms or by mixtures of chemicals.

Yeast, also known as baker's yeast, is my mom's leavening system of choice for pizza crusts (and it is all she knows). Yeast is a living single-celled organism from the fungi family. It ferments (digests) sugars and starches, producing gas (CO_2) and alcohol, which burn off during baking, leaving behind an important component of crust's flavor. Fermentation causes the dough to rise by producing gas bubbles that become trapped in the dough structure. Yeast makes the dough more pliable and therefore easier to handle and form. You can choose from among several different kinds of yeast, including instant, active dry, and compressed. My preference is active-dry yeast. The familiar taste and odor of yeast (alcohol), I believe, is a critical element that must be retained.

Chemical leavening is the creation of gas bubbles by the interaction of a base, typically sodium or potassium bicarbonate, and an acid, often sodium acid pyrophosphate (fast acting) or sodium aluminum phosphate (slow). For more than twenty years, chemical leavening systems have been widely used in industrial products, such as refrigerated dough, batters, and breading. They are now being used in self-rising pizza to make the crusts more tolerant to temperature and time stress prior to cooking. Both a base and an acid are typically used to attain a desired gas bubble size, and timing is key to the entrapment of the bubbles within the dough structure. This is how the basic formula might be written:

Acid + Base $\rightarrow$ Salt + *Gas* (gas = leavening)

A typical chemical reaction to create carbon dioxide gas in baked items looks like this:

$$NaAl(SO_4)_2 + 3\ NaHCO_3 \rightarrow Al(OH)_3 + 2\ Na_2SO_4 + 3\ CO_2$$

Although generating carbon dioxide gas and neutralizing sodium bicarbonate is the primary role of leavening acids, it is important not to forget the other functional effects of leavening acids:

1. They can enhance bread texture, evidenced in the grain, tenderness, and mouthfeel.
2. They enable a precise pH control, which impacts the crumb color and flavor.
3. They improve the consistency of the finished product, as leavening relies on chemical, rather than biological, processes (that is, yeast).

Let us review the available options. First, the various bases that can be used:

- *Sodium bicarbonate* ($NaHCO_3$) is a base ingredient that, when mixed with water and an acid, produces carbon dioxide (CO_2) gas. Sodium bicarbonate is available in a range of particle sizes, which can be used to modify the timing and size of gas bubbles.
- *Potassium bicarbonate* is an expensive alternative to sodium bicarbonate, primarily used for low-sodium applications. In addition, it imparts a less salty taste to the crust.

As stated, the base needs to be combined with a leavening acid to produce carbon dioxide:

- *Sodium aluminum phosphate* (SALP) is a slow-acting acid that reacts primarily during the baking process and provides a springy texture and volume to the crust. It is more stable at room temperature than other leavening acids.
- *Sodium acid pyrophosphate* (SAPP) gives a softer result, producing a finer, less springy, and crunchier texture. It also has less of an aftertaste compared with other acids.
- *Monocalcium phosphate* (MCP) is also known as calcium acid phosphate. It acts fast, responding mainly during the initial leavening. The finished volume of the crust is smaller, and the crust itself has a tighter grain and firmer texture.
- *Sodium aluminum sulfate* (SAS) typically is used in combination with MCP in double-acting baking powder. It has a high neutralizing value and reacts mostly in the baking cycle. It is thus more appropriate to long bake cycles in which cakes are created.

My frustration with using yeast began years ago, when my mother made up her dough and waited (typically at least 2 or 3 hours) for the yeast to generate the rise (gas bubbles) she was looking for. I simply do not have the luxury of time, and chemical leavening offers me the flexibility to create a "mix-and-go" dough.

You may find that these individual leavening ingredients are not easily sourced through conventional suppliers, but the effort to obtain them could be worth it. Alternatively, you may initially want to work with conventional baking powder. Most commercial baking powders are blends of bicarbonate, one or more acid salts, and a starch, which acts as a filler and water-vapor absorber.

As I searched for these individual ingredients, I also obtained samples of encapsulated sodium bicarbonate and an encapsulated blend of sodium bicarbonate and sodium aluminum phosphate (SALP). Encapsulated ingredients are coated forms of common bakery ingredients. Encapsulation is achieved by placing the ingredients in contact with a melted fat under constant and turbulent mixing followed by slow cooling, so as to deposit the fat on the surface of the ingredient. The coatings are vegetable oils with no trans fats and high melting points, typically around 145°F (62°C). Such coating protects the leavening agents from reacting with one another once they are incorporated into the dough, which allows for control over the leavening

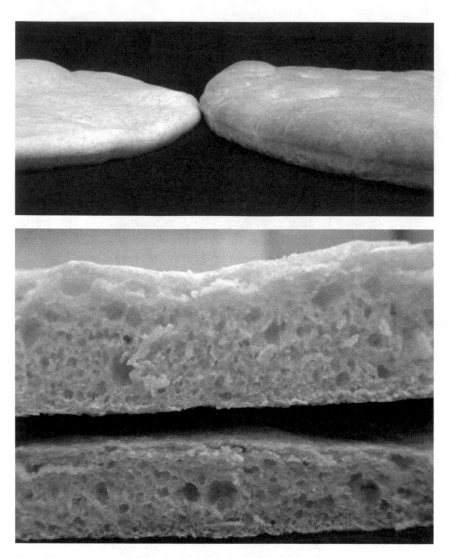

Figure 53 Comparison of pizza dough volume (after baking) obtained by incorporation of encapsulated chemical leavening ingredients in addition to yeast: (*top*) crust made with yeast alone (*left*) and crust made with the addition of encapsulated ingredients (*right*); (*bottom*) cross section of the baked pizza crust. The recipe, using encapsulated leavening (*above*), provides a consistent rise that is tolerant to time and temperature stresses. The crust prepared using yeast is less well developed and more dense (*below*). Both crusts were made up to 24 hours in advance (refrigerated and covered with plastic wrap).

process. The agents will react only after the fat that surrounds them melts away during baking. Encapsulated bakery blends quickly became my secret ingredients. Figure 53 illustrates the effects of the addition of encapsulated leavening ingredients on the volume and appearance of the pizza crust (top crust) compared with yeast raised (bottom crust).

When I baked my first crust with encapsulated chemical leavening—oops—it rose so much that it turned into pizza bread (see figure 53)! But it did not take me long to smooth out the kinks. I modified my mom's recipe and created my own version, a high-tech Old World gourmet pizza. I can make up my dough hours or a day in advance (and keep it refrigerated) and bake my awesome pizza fresh whenever my guests arrive.

Let me share a few more observations:

1. While you can make a terrific crust with only chemical leavenings, in my opinion, adding a bit of yeast contributes greatly to the flavor and aroma (no leavening function, just aroma and flavor), making it just like my mom's pizza.
2. When you apply your sauce, avoid laying on a pool of the stuff. Leave visible gaps so that the heat can overcome the heat-dissipating effects

TOM'S OLD WORLD PIZZA CRUST WITH NEW WORLD INGREDIENTS

6 ounces (180 g) warm water (82°F [28°C]) (34 percent)
10 g vegetable oil (2 percent)
11 ounces (330 g) bread flour (60 percent)
10 g granulated sugar (2 percent)
7 g salt (1 percent)
3 g encapsulated bakery blend (0.5 percent)
2 g yeast (active dry) (0.5 percent)

Make the dough by adding the water and oil to a stand mixer with a dough hook attached; alternatively, you can use a bread machine. Add the flour, sugar, salt, encapsulated bakery blend, and yeast. Mix at a low speed for 10 to 20 minutes.* Occasionally, scrape the sides during the first 5 minutes of mixing.

Dust the pizza stone or tray with corn meal before laying out the crust. The corn meal will allow the dough to slide as it cooks and release more easily. When mixing is completed, roll the dough on a floured surface into the desired diameter. This recipe creates a nice 12-inch (30 cm) pizza. Remember to cover the crust with plastic until you are ready, otherwise it will dry out and yield a tough skin. The crust can now be refrigerated for a minimum of 2 hours (and up to 24 hours) to allow the flour to fully hydrate (chapter 9).† Preheat the oven to 425°F (220°C) and bake for 10 to 15 minutes (cooking time will vary with oven type and topping mass). Chefs know that each pizza is unique, even for the same combination of ingredients. Therefore, browning of the crust and melting of the cheese are important indicators of pizza doneness.

* I prefer 20 minutes to ensure that the encapsulated ingredient is properly incorporated: too little mixing can cause an uneven rise in the crust. Beware of overmixing, either for too long or too fast, as this shears the coating off the encapsulated bakery blend, causing it to start working prematurely and resulting in poor leavening during baking.
† Using the crust too quickly results in a poor rise, which I think has to do with gluten development.

of the sauce and rise evenly through the dough. The impact of this technique on the baking of the crust can be dramatic.

3. Do not allow yourself to fall into a rut. Do not be afraid to vary your crust thickness or vary the type of flour (gluten strengths) you use. Do not just vary your toppings—vary how you apply them. Concentrate certain toppings on certain sections of the pie so that your guests can choose which slice they want to savor.

My pizza (and my mom's) was done in a conventional gas oven at 425°F (220°C) directly on a baking stone. You can bake your pizza in a variety of ovens, including electric deck ovens, conveyor or belt, brick, or wood-fired brick ovens (for a smokier taste). Commercial pizza ovens, including wood-burning ovens, cook at very high temperatures: 480°F (250°C) or higher. My recipe has not been proved under these high cooking temperatures, but it soon will be!

While traditional ingredients can be used to create great dishes, nontraditional ingredients offer chefs, cooks, and avid food lovers the tools to create the same or new dishes that fit with today's hectic schedules, while allowing time for the enjoyment of family and friends.

twenty-nine

EATING IS BELIEVING

LINE HOLLER MIELBY AND MICHAEL BOM FRØST

unsicht-bar was suggested to us in connection with a sensory conference we were attending in Hamburg. We made reservations several weeks in advance. Normally, the restaurant is closed on Mondays, but with thirteen excited Danes as prospective diners, Unsicht-Bar flung wide its doors to us. A nice gesture: our expectations were rising. At the conference, rumors spread that we were going out for a special meal. Thus, we ended up being a party of thirty-four sensory scientists. Sensory scientists will generally go far out of their way for anything having to do with extraordinary food. As for the literally *dark* culinary event we were about to experience, we only knew it was going to open our eyes to new ways of looking at food.

Upon our arrival in the foyer, we were grouped into small parties, and orders for food and drinks were taken. It was explained to us that we could choose from among four menus: beef, poultry, vegetarian, or the surprise menu. Needless to say, the majority selected the surprise menu—why else would we come here? The items on the surprise menu (figure 54) would be disclosed after diners had finished the meal.

Each group of diners entered the dining room together with their waiter and were seated. It was a big room but with good acoustics. The conversation at other tables was audible but not intrusive. Our

Cold cucumber–buttermilk soup with fish
Salad with king prawns, fried salmon, trout caviar, and several antipasti with dill and balsamic vinegar dressing
Ostrich and veal fillet, thyme cream polenta, and balsamico chanterelle
Chocolate fudge with lemon sorbet, vanilla ice cream, and cowberry topping

Figure 54 The surprise menu.

table did not have a table cloth. There were place mats for each of us. The Ikea-savvy among us recognized them instantly. The table arrangement was flexible; there was Velcro at the table edges aiding the stability of our two juxtaposed tables, which created a four top. The waiter arrived with our drinks—wine by the glass. He announced that an amuse-bouche (complimentary appetizer) would shortly follow. We started rolling out the thick paper napkins. One of the diners decided to put his napkin in the neck opening of his shirt. That night he wanted to be particularly sure it did not fall to the floor. The amuse-bouche—prawns on a slice of baguette—arrived on one plate and was placed in the center of the table. Prawns are such a familiar and easy-to-recognize flavor. These had a wonderful soft and juicy texture. When a baguette is good, the crust is so crisp—almost glasslike—that it shatters on the bite. It was an especially good baguette: the textural difference between the crust and the soft crumb was particularly pronounced. Shortly after, our soups arrived. The smoked salmon revealed itself from a distance; therefore, it was easily decoded. The soup bowl had "ears," making it easier to drink the soup. No one at our table was able to recognize the buttermilk in the soup, so it was a surprise when we saw it on the menu afterward.

The second course was a salad, and almost every bite was a surprise. As we were eating from the surprise menu, we were not especially amazed. When the salad course was served, an aroma of clear marine origin enveloped us. We reached a consensus on fish eggs of some kind, but it took many bites before we experienced their texture and taste. Smoked salmon, now part of the salad, again gave itself away from a distance by its smell alone. Fennel was easy to recognize by its aniselike aroma. Normally, the texture of its surface is nothing spectacular. But for us, totally unaware of its presence, it felt remarkably slippery yet not smooth—we could feel the small rims on the surface. The prawns were gigantic, and it was difficult to eat one in a single bite. Remarkably, the dill went unnoticed by several diners, possibly because there was so little of it. Several of us did not notice that there were two different animal species as part of the main course—veal and ostrich. The dessert was literally plate-licking delicious!

At one point, one of the diners had to use the restroom. He called for the waiter. The waiter laid the diner's hands on his shoulders, escorting the diner to the light lock—the light was intensely strong after the total darkness of the restaurant. But it did not matter. The diner had his sight again, and probably anyone going to the restroom was thankful for that! To the waiter, light or dark did not make a difference—he was blind, like the rest of the dining room staff.

It was an extraordinary experience to eat a whole meal in the dark at Unsicht-Bar (*Unsichtbar* is the German word for "invisible"). When we are deprived of one of our key senses, our impression of the world can change radically. In connection with eating, visual input generates expectations about what is to come. In the absence of these expectations, our other sensory impressions of food are dramatically altered. From the first second of darkness, this resulted in a very different impression of the room as the remaining senses sharpened to get a grasp of the surroundings. When first seated, we touched everything, feeling the square table, the round plate, the glasses, and the place mat. The lack of visual stimulation highlighted some sensory properties, like the texture of the baguette and the slipperiness of the outer surface of the fennel. Because appearance is a major factor in the identification of foods with less familiar aromas, tastes, and textures, darkness hinders our ability to decode what we are eating. The setting invited deep discussions about what we perceived and also encouraged behaviors that are otherwise inappropriate in other restaurants—licking your plate, for instance!

Some of us dealt better with the discomfort of visual deprivation than others. Two diners at our table felt quite uneasy at the beginning. Would we be able to leave the room if something went wrong? Things started to feel better when we were served the drinks and the starter. Drinks were given to us using sounds as a directional guide (the waiter clicked his wedding ring on the side of the glass). The experience of the Unsicht-Bar meal was markedly different from that of a meal in a regular restaurant setting. We did not have our sight to help us generate expectations for what we were about to eat. When the aromas from the dish failed to give it away, we had to actually touch the dish before we had any idea what it was. As it happens, our experience eating blind at the Unsicht-Bar was not at all what we expected it would be. We had thought that it would be like being blindfolded in an illuminated room, as in familiar childhood tasting games. The experience is much more powerful than that. Even with our eyes wide open, we still did not see a thing. Obviously, it is difficult to generate expectations about something for which there is no point of reference.

Overall, we found that the conversation was different from the usual chatter. Inevitably, the talk turned to the scene in the movie 9½ *Weeks* in which Mickey Rourke blindfolds Kim Basinger and feeds her all sorts of delicacies. It is a classic scene showing the proximity of food and lovemaking. Other topics we took up were much more serious. It was our feeling that the darkness and the sharing of such an extraordinary experience made the conversation more sincere. And we strongly agreed we had been through something utterly wild, creating a sense of togetherness among us.

This account of our experience at Unsicht-Bar shows that an unusual setting will affect the perception of a meal. It vividly demonstrates that visual input is vital for the generation of sensory expectations, and in the absence of these expectations, many of the sensory impressions change. It highlights that a range of processes we take for granted and effortlessly do in connection with eating are really quite complex.

The Eating Experience

The fact that the eating experience is such a complex, multidisciplinary, and vital phenomenon has resulted in its examination through many different fields of study, including sociology, economics, nutrition, and the arts. Food has always been of great economic concern, and with the growing interest in and attention paid to food culture, health concerns, and the "experience economy," it has become a part of the political agenda. In this buyer's market, food shopping for an increasing segment of the population has changed from a search for inexpensive macronutrients to a search for experiences. A food product provides energy to the body but also conveys meaning through cultural codes. It is an expression of the identity we wish to possess.

The experience of a meal is affected by a variety of factors in addition to the food itself. It is a highly personal process influenced by our physiological and psychological history, attitudes, mood, and other variables. In a restaurant, along with these personal factors, contextual effects, such as the setting, the staff, the service, and the company certainly all contribute to the way the meal is perceived.

Every meal has a host and is dictated by ritual. The host, visible or not, has a major influence on the eating experience of the individual. This influence exhibits itself in a sense of confidence on the part of the host and trust on the part of the guest. Also, the host creates the frame in which the eating takes place. This frame is an important part of the food experience. In some cases, as in fine dining, there is seemingly no limit to the cost and the imagination. The host–guest relationship is of great importance to the dining experience. Good examples of unusual eating places in this respect are the Unsicht-Bar in Berlin and Madeleines Madteater in Copenhagen. At Madeleines, theater, restaurant, and food laboratory are fused together to give an exquisite overall narrative and sensory experience. The waiter's job at both establishments is not just to guide and navigate the guests but also to make them feel comfortable and secure in an unfamiliar situation. According to Jan Krag Jacobsen (2008), the exquisite eating experience is a question of cultural capital. It is a dialectic produced by skilled chefs—but also skilled consumers—qualified to enjoy the experience.

The Role of Expectations

When does the experience of fine dining begin? The experience of a restaurant meal starts long before we sit down to eat. When we read about restaurants, hear other people talk about a specific place, and decide to book a table at a specific establishment, we are already generating expectations about how it will be to dine there. This expectation process continues until we are actually dining. Our expectations influence our experience of the meal. Often, especially in fine dining, our expectations are extremely high, and thus they are at risk of being disconfirmed. The effect of disconfirmed expectations on sensory perception can be accounted for by four different psychological theories: assimilation, contrast, assimilation–contrast, and generalized negativity. Assimilation theory states that an unconfirmed expectation creates a kind of psychological discomfort because it contradicts the consumer's original expectation. Consumers, therefore, try to reduce this discomfort by changing their perceptions to more closely align with their expectations. Thus the *perceived* "product performance"—the meal in a restaurant—assimilates to the *expected* performance. According to contrast theory, disconfirmed expectations will result in an exaggeration of the disparity between expected and actual stimulus properties. For instance, according to this theory, a restaurant's modestly understated advertising will result in higher consumer satisfaction. Conversely, a restaurant that has been presented as better than it actually is will result in a considerable decrease in consumer satisfaction. Contrast theory is, therefore, the reverse of assimilation. Assimilation–contrast theory dictates that there are boundaries inside which our perception assimilates to our expectations, but outside those boundaries, differences between actual perception and expectations will be exaggerated. Finally, generalized negativity predicts that when an experience is not as expected, it is perceived as negative. Most studies in consumer food science support assimilation or assimilation–contrast theory. These theories have never been studied with respect to disconfirmed expectations in restaurant meals, but it is plausible that assimilation–contrast occurs. In the particular case of fine dining, it may well be that the difference between expectations and actual experience need be only very subtle before we see the contrast effect.

Studying the Restaurant Setting's Effect on Consumer Perception

Most sensory consumer studies tend to concentrate on the food itself, with many factors having been found to influence the consumer's responses to the food products. In restaurant settings, other factors affect the perceived quality of the food served. It is possible that the food alone may have less influence on perceived quality than the environmental factors that come into play.

An emerging field within both hospitality science and consumer food science is the study of eating in real-life situations. Research has been done to investigate the effect of a variety of contextual factors on the perception of foods and dishes. In a study conducted in a student cafeteria, Brian Wansink, Koert van Ittersum, and James E. Painter (2005) found that more evocative and descriptive menu names, such as Satin Chocolate Pudding, generated greater positive feedback and higher ratings—the dishes attached to these names were deemed more rich, tasty, and generally appealing than the otherwise straightforwardly named menu items, such as Chocolate Pudding.

Perception in Fine Dining: Challenges and Surprises

Dishes at highly experimental restaurants are different from those at standard restaurants because many of them—besides being delicious—are meant to challenge and surprise the diner. We have examined how different menu-item descriptions affect the diner's perceptions in a restaurant setting. Among other answers, the intention was to discover whether the positive effect of hedonically evocative descriptions of menu items extends to fine dining. In our study, diners were served an eleven-course tasting menu and answered a short questionnaire after each course. Dishes were developed by Torsten Vildgaard, creative sous-chef at Noma in Copenhagen. We used four different descriptions for a single dish: the title only; the title and sensory information; the title and information about the culinary processes; and the title plus hedonically evocative information. Results showed that the kind of information affected the perception of the dish, but there was also an interaction effect with the dish. This indicated that there is a complex interplay between a dish, its presentation, and our perception of it. One dish we served was Brie parfait rolled in rye bread crumble and a rhubarb sherbet. The hedonically evocative presentation was presented as follows:

> Cheese and rhubarb: a delicious and creamy parfait is united with a refreshingly cooling rhubarb sherbet, which assembles in the mouth as pure enjoyment

By contrast, the information about the culinary process was presented in this way:

> Cheese and rhubarb: this dish was frozen at a very low temperature ($-22°F$ [$-30°C$]) and the ice crystals were comminuted using a Pacojet.

Interestingly, people who ate the dish based on the hedonically evocative description liked it the least. Our interpretation is that for these novel dishes, the information about the culinary process better links the raw materials to the served dish, and thus makes it easier for the diners to understand what they are about to eat.

Experimental Studies in Restaurant Settings

With restaurant studies, there are many considerations to take into account. In consumer studies in particular, the experiment is often performed in an artificial eating environment, often a light-, temperature-, and humidity-controlled confined sensory booth. The evaluations of food products under those conditions are stripped of the normal eating context. Therefore, the results are not necessarily reliable predictors of consumer choice in the real world. To achieve valid and reliable data, studies should always be performed in environments as close to the real examples as possible. Scientists seek experimental objectivity and reproducibility of data. This is often accomplished by keeping all factors—other than the experimentally varied—constant and by keeping track of every single detail. Laypersons might label researchers control freaks. However, maintaining complete control in a restaurant setting is impossible. There is simply no such thing as a controlled restaurant environment that simultaneously reflects the true eating situation. We have to live with the challenges of real-life situations, such as diners that make conversation at the table during an experiment.

The mere act of having diners consciously consider their perceptions and give their opinion affects the restaurant experience. This certainly poses research challenges in the field. An alternative approach is the observational study, in which, rather than offer their opinion, diners are merely observed, either by experimenters or video cameras. In the Netherlands, at the Restaurant of the Future, up to two hundred patrons can be monitored by hidden cameras. For more ordinary meals away from home, observational study is definitely a feasible strategy for examining eating behavior. In fine dining, the situation is somewhat different. The motivations behind a guest's choice of dining experience may go well beyond merely eating food to satisfy hunger or choosing dishes that sound mouthwatering. The eating experience in high-end restaurants is as much a narrative for the chef as it is a sensuous voyage of discovery for the diners. Properties of the food, such as perceived novelty, familiarity, and complexity, are central in our appreciation of it. The emotions that the food elicits, such as curiosity or surprise, are necessary for the pleasure obtained from fine dining. Our research is, in part, complete. In future studies in restaurant settings, we hope to gain new insights into the dining experience that we can share and that will further enhance all diners' enjoyment of food.

Acknowledgment

This work is supported financially by the Ministry of Science, Technology, and Innovation through the Danish Research Council for Independent Research Technology and Production Sciences.

Further Reading

Gustafsson, I. B. 2004. "Culinary Arts and Meal Science: A New Scientific Research Discipline." *Food Service Technology* 4, no. 1:9–20.

Jacobsen, Jan Krag. 2008. "The Food and Eating Experience." In *Creating Experiences in the Experience Economy*, edited by Jon Sundbo and Per Darmer, 13–32. Cheltenham, Eng.: Elgar.

Köster, E. P. 2003. "The Psychology of Food Choice: Some Often Encountered Fallacies." *Food Quality and Preference* 14, nos. 5–6:359–373.

Mielby, L. M., and M. B. Frøst. 2010. "Expectation and Surprise in a Molecular Gastronomic Meal." *Food Quality and Preference* 21, no. 2:213–224.

Wansink, B., K. van Ittersum, and J. E. Painter. 2005. "How Descriptive Food Names Bias Sensory Perceptions in Restaurants." *Food Quality and Preference* 16, no. 5:393–400.

Wurgaft, B. 2008. "Economy, Gastronomy, and the Guilt of the Fancy Meal." *Gastronomica* 8, no. 2:55–59.

thirty

MOLECULAR GASTRONOMY IS A SCIENTIFIC ACTIVITY

HERVÉ THIS

many terms, such as *molecular cooking, molecular cuisine, science-based cooking, scientific cooking* (an oxymoron), and *culinary chemistry* (another oxymoron), contain references to science and to cooking. There is much confusion as to their meanings, however, and clear definitions are needed.

Science aims to look for the mechanisms of phenomena, whereas cooking is all about preparing food. In other words, science and cooking can never truly meet, and only technology or education can bring results from applied science into the kitchen. For technology transfers, scientific results are needed, which means the chemical and physical phenomena of food processing must be analyzed. This is the purview of the scientific discipline that was created in 1988 under the name molecular gastronomy. On the application side, "molecular cuisine," which is defined as cooking with new tools, new ingredients, and new methods, has been around for about three decades. It is very fashionable today, but as we shall see, the future of cooking might actually be "note-by-note cuisine," which takes molecular cooking to a whole new level.

In this chapter, I first explain the idea behind this next trend and then explore a menu containing two note-by-note food items. In the process, I define various terms used at the interface of science and cooking. A glossary at the end of the chapter lists these definitions for future reference.

Let us begin with a clear definition of note-by-note cuisine: it is a culinary trend in which no plant (vegetables, fruits) or animal (meat, fish) tissues are used, because these traditional food ingredients are mixtures of compounds giving poor control to the cook. Instead, note-by-note cuisine makes use of "pure" compounds in order to build all aspects of dishes: taste, odor, color, texture, and so on.

Now, let us make comparisons in order to better appreciate the interest of the proposal. First, let us observe that the philosophy of note-by-note cuisine mirrors music, specifically the kind of music that uses synthesizers instead of traditional instruments. Music can be created by mixing sound waves of "pure" frequency. This music can of course reproduce the effect of flutes, violins, and pianos, with their specific frequency ranges and characteristic sounds. However, it is quite useless to reproduce, and artists would prefer using the new tool to produce new sounds and music. In the analogy, the pure compounds in food resemble the pure frequencies in music, while particular plant- and animal-based materials—each of them with their specific sensory components—resemble the traditional instruments.

The number of new possibilities given by note-by-note cuisine is tremendously greater than with traditional food ingredients. Many interesting questions are raised by this new practice that began practically in 2002 (whereas the theoretical proposal was done by me and Nicholas Kurti in an article published in 1994). It was finally realized as a sauce—the Wöhler sauce—by Pierre Gagnaire at Sketch, his restaurant in London. The next note-by-note dish was an entrée on Pierre's Art and Science menu in Hong Kong on April 24, 2009. More recently, French chefs have proposed note-by-note dishes or whole menus: Hubert Maetz and Aline Kuentz in Strasbourg (May 2010), the chefs of the Cordon Bleu School in Paris (October 2010), and Jean-Pierre Biffi and his team at Potel & Chabot catering company in Paris (January 2011, April 2011).

An Art and Science Menu

Given the complexity and length of the Art and Science menu served by Pierre Gagnaire, I do not attempt to describe all the dishes in detail (figure 55). Instead, I outline the basic idea and knowledge that went into creating each of the menu items. Let us say first that only one dish is fully note-by-note (the first); the others are examples of molecular cuisine.

Note by Note N°1

Foie gras Chantilly/avocado/black radish

Cucumber Liebig on thin slices of Mediterranean sea
bream perfumed with dry bonito
Shellfish and acidulated mixed herbs

Abbé Nollet leaves, Turkish morels with vin jaune
from the Jura, and spring onion
Semolina with turmeric

Kientzheim butter on a fillet of sole veiled with white corn flour
Spring soup and parmesan cheese

Abstract

Slow-cooked chicken consommé simmered for 3 days to obtain its optimum taste
Bresse poularde breast Shimizu
Coconut milk and lemongrass panna cotta

Polyphenol sauce seasoning a blue lobster fricassée poached in a noisette butter
Potato Tamy and dry black olives

Turnips Shitao

Wurtz: with Roquefort cheese terrine and milk jelly with green curry

Raspberry puff pastry and jelly: vanilla diplomat cream and cappuccino

Wind crystal: strawberry ice cream and grapefruit marmalade

Gibbs: passion fruit and honey emulsified with olive oil in a goat cheese parfait

Brown sable roasted: coffee jelly and unctuous whisky ganache

Figure 55 The Art and Science menu, which included the first note-by-note dish in the history of cooking.

• *Note by Note, Number 1.* In this dish, the major compounds are water, citric acid, milk proteins, maltitol, sodium alginate, calcium lactate, glucose, sodium chloride, and ethanol (figure 56). It includes alginate "pearls" with a liquid core, dispersed in a sherbet with layers of crispy "peligot" (that is, glucose caramel) disks. The flavor of the dish? The question is frequently asked, but how could an answer be given? Imagine that you are blind, and you ask a friend what the color blue looks like. How could he explain? This is clearly a big issue for note-by-note cuisine, as it was for synthesizers in music when new sounds were produced.

Figure 56 The first note-by-note dish: Note by Note, Number 1.

• *Foie gras Chantilly/avocado/black radish.* Another issue with Pierre Gagnaire's dishes is that their descriptions are too lengthy. Instead of losing time in details, let us say only that the dish contains an avocado preparation, black radishes, foie gras, and dozens of other ingredients. The interesting part relies on the technology transfer: foie gras Chantilly. This is a generalization of whipped cream obtained by first making an oil-in-water emulsion from foie gras (mainly fat) and water, and then whipping it as we cool it so that fat crystallization occurs. In doing so, the fat droplets partially coalesce and crystallize, thereby creating a crystallized fat network that entraps the air bubbles. In whipped cream (called Chantilly when sugar is present), the fat is from milk. Taking this concept of whipped cream by using fat crystals to stabilize air bubbles, I proposed using, instead of milk fat, other fats and bases, like chocolate to make chocolate Chantilly or cheese for cheese Chantilly.

• *Cucumber Liebig on thin slices of Mediterranean sea bream perfumed with dry bonito; shellfish and acidulated mixed herbs.* Here the technology innovation is the "liebig," a new culinary preparation (molecular cuisine) attributed to the German chemist Justus von Liebig. Liebig preparations are jellified oil-in-water emulsions that are obtained by first making an emulsion using gelatin, oil, and an aqueous solution (tea, coffee, wine, or stock, for example). When the emulsion cools, the gelatin molecules form a gel in which oil droplets are trapped. Note that this gel is a physical gel. When the temperature is increased, the gel melts and turns back into an oil-in-water emulsion.

• *Abbé Nollet leaves; Turkish morels with vin jaune from the Jura, and spring onion; semolina with turmeric.* In this dish, the idea is to produce artificial salads based on the observation that sheets are wonderful technological objects that fold easily when a force is applied perpendicularly to their plane and offer great resistance when pressed along the plane. During the chewing of the sheets, the two positions occur at random; this difference is important because our senses and brains recognize contrasts. Nature provides an infinite array of sheetlike materials, but cooks have almost no control over their aroma, taste, or consistency. This is why we suggested making "artificial" sheets, instead of plant leaves. Making sheets is not difficult: they can be produced with a great variety of compounds, such as by pouring an aqueous solution mixed with gelatin on a flat plate or by heating a liquid containing coagulating proteins. Even salad dressing can become a sheet. This dish was named after Jean-Antoine Nollet, a pioneer of the science of electricity (figure 57).

Figure 57 The Nollet salad.

• *Kientzheim butter on a fillet of sole veiled with white corn flour; spring soup and parmesan cheese.* Here the new product is called a kientzheim, after a small Alsatian village. The idea is to make a mayonnaise using brown butter instead of oil to obtain a new flavor profile. The brown butter should not be too hot (below 140°F [61°C]) or the egg will coagulate during the emulsification process, yielding a traditional béarnaise or hollandaise.

• *Abstract.* Here, neither science nor technology plays a role—only aesthetics. Many years ago, I suggested that artistic trends such as impressionism, cubism, and abstract expressionism found in other art forms (for example, music, literature, painting) could be transferred to the culinary arena. As Wassily Kandinsky ([1923] 1979) explained well in abstract painting, the idea is to create emotion without "reproducing" familiar forms such as houses, people, plants, or landscapes. Using the same approach here, the idea is to produce a dish that is "good" but in which no "flavor form" (that is, traditional food taste or smell) can be recognized. This particular dish was based on the flavor of egg yolk, ham, and cardamom—and nobody was able to guess it!

• *Chicken consommé cooked for 3 days to obtain its optimum taste; Bresse poularde breast Shimizu; coconut milk and lemongrass panna cotta.* Here the main word is *taste*, which we should distinguish from flavor, the latter being the comprehensive sensation one experiences when eating. Taste is only one component of flavor, and it stems from the compounds that dissolve in the saliva before reaching taste receptors from cells in the taste buds. The important idea behind this dish is that proteins are hydrolyzed through long-heat treatment (cooking) so that amino acids are formed. As amino acids have flavor, stocks can have a very strong taste when heated for an extended period. In the case of collagen, the main amino acids formed after cooking this long—way past the formation of gelatin—are glycine, proline, and hydroxyproline.

• *Polyphenol sauce seasoning a blue lobster fricassee poached in a noisette butter; potato Tamy and dry black olives.* This was an earlier test to make a fully synthetic note-by-note sauce. The name given to this polyphenol sauce was Wöhler sauce, in honor of the German chemist Friedrich Wöhler. The phenolics were extracted from grape juice (in this particular case, the syrah grape) using reverse osmosis, a physical method frequently used for water purification (figure 58).

• *Turnips Shitao.* The idea here was to use capillary effects. A piece of porous material can soak up any liquid under the right

Figure 58 Lobster with Wöhler sauce.

conditions. This effect is useful for introducing a particular flavor inside a material, such as plant or animal tissue. The name of the seventeenth-century Chinese painter Shitao was given to dishes using this effect. French chef Pierre Gagnaire had the idea to "paint the plate" using modified plant tissue (from turnips), which lends some visual art to culinary art.

• *Würtz with Roquefort cheese terrine and milk jelly with green curry.* Charles Adolphe Würtz, one of the important chemists of the nineteenth century, promoted atomic theory but also discovered chemical compounds such as amines, glycol, and aldol. He showed that glycerin is a tri-alcohol and published many methods of organic synthesis. I gave his name to dishes in which a jellified foam is obtained by whipping gelatin in water. Gelatin, a protein, enables the formation of large amounts of foam (chapter 14). The interesting part comes when the liquid foam spontaneously transforms into a solid foam by the action of cooling, which physically, and reversibly, cross-links gelatin strands.

• *Raspberry puff pastry and jelly: vanilla diplomat cream and cappuccino.* Since making puff pastry requires water, why not substitute a flavored solution (with raspberry flavor)? The leavening action obtained during baking is based on water evaporation, so we maximized this effect: we increased the quantity of water in the dough by emulsifying the water into the fat (in this case, butter with 40 percent instead of 20 percent water). A jelly was added to the pastry, while this first piece was accompanied by a vanilla diplomat cream and a vanilla cappuccino.

• *Strawberry ice cream and grapefruit wind crystals.* Here the innovation is the wind crystal, based on the following idea: theoretically, one can get about 35 cubic feet (1 m³) of whipped egg white with only one egg white. The volume of foam is limited only by the quantity of water in the egg white; that is, there is enough protein in one egg white to yield 35 cubic feet of foam. As demonstrated in chapter 15, foam creation is inhibited only at very high dilutions of egg white with water. Hence, the idea of adding an aqueous solution (fruit juice, wine, or stock, for example) to an egg white to obtain a large quantity of very light and flavorful foam has merit. Add some sugar—sweet or not (as described in chapter 24)—pipe it, and bake it, and you will end up with a tasty, glassy meringue: a wind crystal (figure 59).

• *Gibbs: passion fruit and honey emulsified with olive oil in a goat cheese parfait.* The dish has many elements, one of which is the food preparation that I call gibbs. This is named after the American physical chemist Josiah Willard Gibbs, one of the pioneers of thermodynamics. With this dish, an emulsion is trapped in a *permanent chemical* gel as opposed to a reversible physical gel, as in the case of a liebig preparation. The recipe is simple. Just whip oil in an egg white to make an oil-in-water emulsion. Then heat the emulsion in a microwave oven for a few seconds, until the preparation expands

Figure 59 Wind crystals as part of a dessert.

and cooks. The coagulation of the egg-white proteins traps the oil droplets, resulting in a gibbs.

• *Roasted brown sable filled with coffee jelly and covered in whisky ganache.* The idea is to "cook" the flour before making the cookie dough, which impairs the development of the gluten network responsible for cookie hardness. Cooking the flour is achieved by roasting it in an oven at 350°F (175°C) for about 15 minutes or simply cooking it in a pan until it slightly browns. In this way, a perfectly sandy texture is obtained—hence the name shortening. A sandy consistency is not the only goal in this process. A dark brown color and rich flavor are also desired, and these are achieved in the roasting process via various reactions, such as pyrolysis, hydrolysis, oxidation, and the Maillard reaction (chapter 13). For reasons that remain to be investigated, the roasted flour can be described as having a flavor that ranges between mushroom and chocolate, depending on the roasting conditions.

We hope that this menu will inspire the use of a scientific understanding for the good of cookery. Let us celebrate knowledge in general and chemistry in particular!

A Glossary for Future Reference

What follows is a glossary of terms that may be useful when working within the realm of culinary technology. This part is not superfluous: the great French chemist Antoine-Laurent de Lavoisier rightly wrote that science means studying phenomena and that this study needs the right words. One cannot improve science without improving the language, and vice versa. This was indeed also the idea of the British physical chemist Michael Faraday, who introduced words such as *anode* and *cathode*. This is why we need to clarify terms to better understand the relationship between cookery, or cuisine, and molecular gastronomy.

• *Science.* There is much debate about the meaning of *science*, particularly among epistemologists. But many scientists would be happy to accept the idea that science is the activity of looking for the mechanisms of phenomena, or trying to picture how things work, using what is called the experimental method or hypothetico-deductive method or simply the scientific method. A similar viewpoint has been expressed by people like Galileo and the philosopher Karl Popper ([1934] 1992), for example. The method involves (1) observing a phenomenon, (2) quantitatively characterizing the phenomenon (and producing data), (3) synthesizing the data into laws, (4) proposing theories (that is, lists of mechanisms explaining the laws), (5) making predictions from theories, (6) testing experimentally the theoretical predictions in the hope that they will be refuted so that the theories can be improved, and (7) going on forever with steps 2 through 6. It can be seen from this description that science will never be "in the kitchen," because science produces knowledge (mechanisms of phenomena) and not dishes! Hence the question: What can science and cooking have in common? I have the feeling that the answer is nothing.

• *Cooking.* People forget that cooking was, is, and will always remain the activity of preparing dishes; it can be a craft or an art, but dishes will be produced for human consumption. Recognizing the nature of cooking is of the utmost importance because it allows us to distinguish "food" from "dishes."

• *Food.* What is food? Dictionaries define it as "any substance that can give to living beings the elements necessary for their growth or for their preservation." According to this definition, raw plant or animal tissues should be considered food just as prepared dishes are, but this is confusing and does not do justice to the thought process of putting a meal together. Human beings seldom eat untransformed tissues or raw natural products. Raw materials are modified, and chemical and physical changes determine food's final composition and structure, as well as its bioactivity (sensory effects, nutritional value, possible toxic effects, and so on). Plant or animal tissues are at least washed, cut, and thermally processed. Generally, cooks (even in the food industry [the difference between home, restaurants, and food factories is generally a question of scale, not the nature of the products]) devote themselves to maintaining the safety of food ingredients, and in the cooking process, cooks change foods' consistency and flavor. Even for a simple carrot salad, there is a big difference between the raw product in the field and its presentation on the plate.

• *Technique.* The word *technique* comes from the Greek word *techne*, which means "doing." Technical activities in the kitchen produce results in the form of prepared dishes. And this is why many "chemical arts"—drug making, cosmetics production, painting and dye production, candle making, and of course, cooking—are technical activities. In spite of the adjective *chemical*, which has a long history, cooking is not really a chemical art because chemistry is a science (rather than a technique or a technology).

• *Technology.* The word *technology* should be clear because its etymology is from *techne* and *logos* (study). Technology is the activity of improving technique, with or without scientific results. Alas, too frequently technology is called applied science, a faulty expression because the application of science is no longer science, but rather technology. Science does not care about its application. In cooking, there is one specific technology, which I propose to call simply culinary technology.

• *Chemistry.* Chemistry is the science that studies the mechanisms of atom rearrangement with simultaneously occurring changes in electron pathways that are shared between these atoms. The changes may occur in molecules or in other structures constructed with atoms. Some of the atom rearrangement occurs during culinary processes, but it does not mean that cooks are chemists. A kitchen can be a laboratory only if we stick to the etymology of the word *labor*: a room where people work.

• *Physics.* Physics is the science that studies the mechanisms of change in space–time properties of matter at any possible scale (from meters to nanometers and beyond). Changes such as these occur during culinary processes. Although cooks are not physicists, they can improve their techniques using results from the physical sciences.

• *Gastronomy.* Here again, there is much confusion: many people think that gastronomy is cooking for the wealthy or with costly ingredients. On the contrary, the word *gastronomy* was introduced in French in 1801 by the poet Joseph Berchoux and then popularized by Jean-Anthelme Brillat-Savarin by means of his literary masterpiece *Physiologie du goût* (*The Physiology of Taste*), published in 1825. In this book, he defined gastronomy as "the reasoned knowledge concerning all aspects of food." Gastronomy is knowledge, then, and for the sake of proper thinking, we should avoid using expressions such as "gastronomic restaurant."

• *Art.* The meaning of art has changed extensively with time, and I am not able to define it in a few words. However, today, art is more or less an activity of creating emotions in connection with beauty. In cooking, *beautiful* means "good to eat." This is really too short a description for such a complicated term. However, shouldn't we keep the idea that some cooks are more like technicians and others, artists? Food for thought . . .

Further Reading

Bocuse, Paul. 1976. *La cuisine du marché*. Paris: Flammarion.

Brillat-Savarin, Jean-Anthelme. 1825. *Physiologie du goût*. Paris: Brillat-Savarin.

——. 1971. *The Physiology of Taste, or, Meditations on Transcendental Gastronomy*. Translated by M. F. K. Fisher. New York: Knopf.

Kandinsky, Wassily. (1923) 1979. *Point and Line to Plane*. Translated by Howard Dearstyne and Hilla Rebay. New York: Dover.

Popper, Karl. (1934) 1992. *The Logic of Scientific Discovery*. London: Routledge.

This, Hervé. 2006a. "Cooking in Schools, Cooking in Universities." *Comprehensive Reviews in Food Science and Food Safety* 5, no. 3:48–50.

——. 2006b. *Molecular Gastronomy: Exploring the Science of Flavor*. Translated by Malcolm DeBevoise. New York: Columbia University Press.

This, Hervé, and Pierre Gagnaire. 2006. *La cuisine, c'est de l'amour, de l'art, de la technique*. Paris: Odile Jacob.

——. 2008. *Cooking: The Quintessential Art*. Translated by M. B. DeBevoise. Berkeley: University of California Press.

This, Hervé, and Nicholas Kurti. 1994. "Physics and Chemistry in the Kitchen." *Scientific American*, April, 44–50.

thirty-one

THE PLEASURE OF EATING

The Integration of Multiple Senses

JUAN-CARLOS ARBOLEYA, DANIEL LASA, OSWALDO OLIVA,

JAVIER VERGARA, AND ANDONI LUIS-ADURIZ

some years ago, an interesting collaboration was conceived between the Spanish restaurant Mugaritz, near the city of San Sebastián, and scientists from Azti-Tecnalia Food Research Institute in Bilbao to investigate some of the factors that make the experience of eating so unique.

The collaboration was motivated by a desire to better understand the events surrounding our everyday eating activity. Our main objective was to examine exactly what goes into the construction of pleasurable culinary experiences. But what is pleasure? Although the challenge of exploring the concept of pleasure from a theoretical and practical point of view presented itself as an attractive project, it turned out to be difficult to realize in practice. Nevertheless, the proposition remains and our research continues!

Reflections of a Cook

The belief that the pleasure experienced during a good meal comes only from the response to sensorial stimuli is an insufficient, even naive, view of what is really going on. This is a fact of which cooks are becoming more and more aware.

We all have experienced disappointment in a restaurant at some point, especially when we do not get complete satisfaction at an establishment praised for it exceptional gastronomic virtues. Sometimes this happens even when high-quality ingredients and the skill displayed by the chef perfectly fulfill our expectations. We can make sure that our senses are particularly excited and alert, that we are open and prepared for the coming experience, and still disappointment can occur. Why it is that even under favorable conditions, satisfaction can elude us? Where does the process go wrong? It turns out that the experiential aspect, the acclimating to a new experience and then the bonding to it, involves feelings—the expectations, beliefs, and compromises that emerge between the diner and the dining experience—that develop on an emotional level. And these feelings have the power to shape our own sensorial perception.

Perhaps it is possible to analyze disappointment objectively by assessing the quality of the experience. We might look at specific parameters, like the quality of the ingredients and the service, the level of creative skill, and so on. Undoubtedly, the creative-skill parameter would be very difficult to rate because experiencing pleasure while eating a dish is not the result of just culinary skill but of many creative aspects. In fact, pleasure is a response to a range of different motivations and circumstances.

There are some aspects to culinary pleasure that, apart from being captured through our senses, are intimately connected to emotions, intellect, and even imagination. It is in this ambiguity of concepts and in this abstract space, where rules are not arbitrary, that the senses are met by pleasure.

These ideas are easily illustrated in a Japanese ceremony called *senchakai*. The ceremony consists of an elaborate ritual that takes place in a traditional Japanese teahouse (*jingaisitsu*). The master of the ceremony (*tsukuda-sensei*) leads the attendants through the ritual, which is divided into four parts, each of which takes place in a different room. The *tsukuda-sensei* presents a series of subtleties, degustations, and reflections meant to illustrate an "idea"—considered the subject of the day—which the attendants must then guess at. The subject can be an emotion, a state of mind, a word, a feeling, or a concept, and every detail in the *senchakai* ceremony is connected to this idea.

The first room, the welcome room, induces the attendants to acquire a contemplative mood through the observation of painting, sculpture, and other elements that are aligned to the subject of the day. The evocations coming from this contemplation represent the starting point of the sensorial experience, which in this case is manifested by a display of art. The taking in of the visual stimuli, together with the deep reflections inspired by the *tsukuda-sensei*, give the attendants the necessary clues to guess the subject of the day. Knowing the subject of the day is very important. This is because every aspect of the coming ceremony, down to the smallest detail, will be strongly related to it. The *tsukuda-sensei* observes and determines whether the attendants fully understand the ceremony.

In the second room, the *tsukuda-sensei* performs a tea tasting with concentrated teas. He plays with the types of tea, their origin, the infusion time, and the water temperature to induce a wide range of organoleptic sensations. It is possible to find a connection with the subject of the day by means of flavor, texture, and persistence of taste combined with the comments of the master. In this room, one may acquire a much wider comprehension of the idea that inspires the ceremony.

The third room has a clue related to the subject of the day in the *tokonoma* (alcove). It is very similar to the second room, only this time, *kaiseki*, a traditional multicourse Japanese dinner, is served. Food now becomes a way to understand the "idea" that inspires the ceremony and hence the subject of the day.

The experience finishes in the smaller fourth room, where dessert and light teas are served and the attendants reflect on the experience with the master of the ceremony.

The *senchakai* ceremony, far from establishing a close set of guidelines, opens a full field for creating individual interpretations. Every gesture performed during the ceremony is meant to build a bridge between the senses and emotions, such that our senses somehow aspire to reach the level of our emotions and motivations. Thus, the line between reality and imagination vanishes. Our emotions shape our senses, almost completely affecting how we perceive beauty, harmony, and even displeasure or suffering.

Therefore, how objective are the senses that influence our perception? How do other aspects—such as suggestion, evocation, imagination, knowledge, culture, state of mind, and the interaction between the senses—interfere and influence our perception? Can we play with these aspects and conditions to guarantee our guests a pleasurable and unique experience? How can cooks modify stimuli to create a positive experience that leads our senses and emotions in a particular direction? Can science help control the experience of pleasure derived from eating?

Reflections of a Scientist

From an exclusively physiological point of view, we eat because we have the impulse or drive to obtain energy and essential nutrients. In other words, it could be said that the pleasure of eating comes from the wish to satisfy a basic need of existence. Subsequently, this pleasure could correspond to the different ways there are to achieve such satisfaction. One of these ways—gathering for a meal—has been identified as the first example of human socializing. So it can be said that the concept of pleasure shows a clear social component, which can be considered as basic a need as food itself.

Moreover, pleasure generated by eating implies by itself a personal interpretation of what is being consumed. And this is because the perception and the interpretation of food are not generated by the food itself but from parts of it. For instance, only a few compounds of the food and the specific combinations among them control the perception of the flavor as a whole. The senses are responsible for capturing the stimuli that emerge during the act of eating or drinking, but they perceive only the stimuli for which they have been designed. This systematic but selective choice of stimuli converts the senses into "reductive filters" of reality. Once food has been ingested, and while nutrients play their nutritional roles, the senses send a message to the brain (information) and a mental image of reality is created. Some researchers argue that an increase in the unpredictability of that information provokes an increase in the feeling of surprise, the stimulus level, and thus pleasure. Therefore, the amount of information is directly related to the level of unpredictability or surprise. Nevertheless, this surprise could easily be unpleasant, due to the fact that, among other factors, our preferences for food have already been determined by an associative learning in early life. Until the age of three or four, we do not develop explicit memories (infantile amnesia), but we grasp a vast amount of information related to emotions. Hence, a baby flourishes on facts and learns flavors early on. This is a key step in our acquiring healthy and pleasurable eating habits in the future, although our food preferences could be limited depending on later experiences.

Keeping in mind these determining factors and how they relate to surprise and associative learning in early life, one can be immersed in the sensorial aspects of food. The senses are the physiological manifestations of perception. Traditionally, we tend to view the five senses—sight, hearing, taste, smell, and touch—as independent from one another. But this viewpoint does not factor in the pleasure of eating. First, it has been proven that the number of senses in human beings exceeds the standard five. Second, the senses are inextricably interconnected. The human brain uses very specific rules to process the information provided by the senses. It collects the sensorial signals related to food, which if taken individually would be very weak. But when collected in combination, the senses provide a sensorial perception far stronger than the simple addition of the individual sensorial inputs. This concept means that multisensorial integration takes place only for those combinations of flavors that appear together in the food we eat. The appearance of food generates expectations about food flavor that can enhance the flavor and thus our eating pleasure. For example, investigations have reported that changes in the color of foodstuffs can cause a change in how flavor is perceived. Likewise, the sound emanating from food can influence the perception of flavor: *crunchy food tastes better* (for a discussion of crispy–crunchy foods, see chapter 2). Ambient sound effects can have considerable influence as well: *fish can taste more pleasant by listening to the sound of the sea.*

Considering all the aforementioned factors, a good understanding of the rules that govern multisensorial integration or multisensorial enhancement has the potential to facilitate the design of ever more pleasurable food experiences. This design should be focused on the food product and the eating environment.

Concerning any particular food item, food enjoyment is governed by a combination of multiple perceptions, including visual appearance, flavor, taste, and texture. Regardless of the obvious importance of flavor and taste, a good understanding of the structural changes in the food product that occur in the mouth can clearly fill an important gap in relating food structure to sensorial perception. For instance, the intrinsic self-assembly abilities of food ingredients can easily control flavor release so that many physicochemical factors linked to aroma affect the overall perception of food. Research into food microstructure as it relates to multisensorial perception must be therefore one of the key factors in the design of pleasant food products.

The sensorial perception of texture depends on the food's microstructure as well as its composition and physical state. For instance, flavor and texture, two of the most important attributes of a sauce,

Figure 60 Presentation of final dishes using a gelatin-based sauce: baby squid baked on a grill fired with vine shoots with vegetarian salted fresh "lard" and reduced cream of "begihandi" squid. (Photograph courtesy of José Luis López de Zubiria [Mugaritz])

are deeply related to the use of butter as a base in traditional cuisine. The use of butterfat aids in the formation of an emulsion that renders a thicker texture, more complex and longer-lasting flavor release, and brighter appearance of the sauce. Replacing butter with gelatin in a sauce leads to some interesting changes. The microstructure of the sauce is completely different, and as a consequence, its palatability and the intensity of the flavors change as well. Also, the presence of gelatin fosters an increase in color intensity and brightness, whereas a fat-made sauce tends toward a grayish tonality (figure 60). Likewise, the microstructure of aerated dishes influences sensorial perception, and the consumption of aerated dishes initially induces a state of satiation despite the reduction in product intake. In addition, the aromas of aerated products are more rapidly released into the oral cavity during the mastication process, changing the sensorial experience. Figure 61 shows a cold soufflé with a high level of aeration that significantly enhances the perception of taste (for a discussion of the role of gases in aerated foods, see chapter 14).

The appreciation of texture in food can involve input from one or more stimuli, including visual, auditory, touch, and kinesthetic. Kinesthesia is different from the sense of touch. It is the perception of the body and its position in space as well as the ability to feel the actions of the limbs—through receptors in muscles—in relation to the body. Touch involves perception via the skin. Although sight

Figure 61 The contrasting sensations of lightness and richness in a dish: cold nut soufflé on "sand" prepared from wall fern dressed with licorice extract. (Photograph courtesy of José Luis López de Zubiria [Mugaritz])

and touch provide valuable information, oral processing is key to the perception of texture in the appreciation of food. Processing food during mastication takes place in the oral cavity. It starts with the first bite, and from there, food is processed into a slippery bolus that can be easily swallowed. The oral processing of food involves a series of forces at the tongue's surface produced by tongue movements, which are picked up by the tactile receptors and then translated into tactile sensations. During this process, saliva plays an important role as it dilutes the food, influences the release of flavor, and allows the salivary enzyme amylase to reach the starch present in the food and break it down. Recently, important progress is being made in the understanding of the in-mouth sensorial perception of food materials by considering structural changes to the food that occur in the mouth.

It seems clear that the perception of texture is a dynamic process based on continuous food breakdown. Different studies have been carried out to quantify sensorial properties of food based on physical and chemical measurements. The relation between material and sensory properties can be clearly seen in the perception of crispness (as discussed in chapters 2 and 21). This characteristic is affected by some components (in particular the water content and the types of carbohydrates and proteins present), mechanical properties, the physical state, and the morphology of the food, which affect sound emission and strength requirements during food crushing. Another example is the perception of creaminess in an emulsion, which is enhanced by either increasing the viscosity or by reducing emulsion stability rather than by adding a larger amount of oil. Increased viscosity promotes coalescence of the droplets and will facilitate the deposition of the fat onto the tongue surface. It seems clear, then, that improved knowledge of the molecular and physicochemical dynamic changes in food structure that occur during mastication facilitate the rational design and fabrication of food, thus increasing both wellness and pleasure.

Synergy Between Cooks and Scientists

The previous reflections give us useful guidelines to help enhance the pleasure we derive at the table. Science is a tool that can be used to control sensorial stimuli in order to generate pleasure when eating. This tool can be employed together with the imagination of the cooks, who are capable of brilliantly integrating scientific aspects (food chemistry and physics; neuroscience and perception) with the psychological, sociological, and cultural ones. It would not be a surprise to see in the near future chefs designing new dishes—by applying new concepts from multisensory perception—that more effectively stimulate the senses. Only in the past few years have psychologists and cognitive neuroscientists started to understand how the human brain combines the different sensory signals during eating. These new insights could help create new eating experiences that are more effective in stimulating the mind.

Furthermore, another aspect justifies this unique collaboration. The investment of time and resources in research and development is a luxury that not many chefs can afford. A restaurant that chooses to avail itself of the resources generated by a wide range of research could face significant financial repercussions. Maybe this is one of the reasons chefs choose to work highly empirically and on their own. Food research institutes have the resources that make it possible to carry out numerous investigations on topics relevant to chefs; however, these institutes often have no real idea how to apply scientific results to applications in the kitchen. The synergy between scientist and cook appears to be complex and perhaps forced at first sight, but it could be a very natural process that promises great advantages to scientist and cook alike. A work in tandem would make it possible to align senses and emotions toward the goal of delighting diners with ever greater eating pleasure.

Transformations in the Kitchen

The kitchen—be it a place to satisfy our most basic needs or an instrument for re-creating experiences in a more emotional territory—seeks tools to adapt to the social, economic, and cultural context of today. To achieve this adaptation, it is essential to take a flexible stand and to update concepts that are no longer sufficient to modern reality.

We are very lucky to live in a time of extraordinary developments and advances in the information sciences, as well as one in which there is a general acceptance of synergies and collaborations between knowledge disciplines. The organization of multidisciplinary research teams that seek more complete answers to key scientific problems is an increasingly occurring phenomenon. Our team of cooks and scientists has paid special attention to the fact that consumers want to experience the pleasure of food but also want nutritional value and fewer calories. One interesting dish design was based on the idea that aerated products could potentially achieve a reduction in caloric density and induce satiety through novel gastronomic structures. Given the same volume of food, it may be advantageous to replace more energy-dense ingredients with equally satisfying flavor components. It has been reported that satiety signals differ as the meal moves through the gut, which include oral (taste and texture), gastric, and intestinal factors. Our prestigious chefs propose another aspect of satiety: visual. They firmly believe that feeling of fullness starts before the food is eaten, at the point at which the food is first viewed by the diner.

Initial attempts at collaboration between scientific disciplines and haute cuisine necessitate appropriate tools for communication that permit the mutual understanding of needs and capabilities. In addition, results need to be transmitted between the two fields in meaningful and productive ways. Science can provide many of the answers to the eternal questions the kitchen hides behind. To identify the important topics, it is indispensable to promote a culture of joint collaboration and research so that the empirical hypotheses cooks hold as absolute truth can be properly evaluated. We now have the capacity to definitively confirm some of the empirical theories held by cooks, but we can also investigate less tested and less scientifically grounded theories.

For all the reasons mentioned, the creation of a common language as a tool for dialogue is essential. The basic vocabulary of the kitchen is built on top of a particular technical glossary of terms (formulated from a practical standpoint). It reaches a superior precision and richness, however, if it is complemented with the appropriate scientific language—a language that conveys empirical concepts together with data and information. This language should be as simple as possible. At the same time, the culinary practice must transfer the sensorial and emotional experience to the scientific realm. These information bridges between cooking and science offer rich sources of data from which haute cuisine can draw and evolve. The culinary advances made in research and science put at the cook's disposal the tools with which to enhance our eating pleasure like never before. A field for exploration and creative development has opened wide!

Acknowledgments

This work was financial supported by the Ministry of Environment, Land Planning, Agriculture, and Fisheries of the Basque government. The authors would like to acknowledge Gemma Garcia for her useful comments and deep discussions during the writing of this chapter.

Further Reading

Arboleya, J.-C., I. Olabarrieta, A. Luis-Aduriz, D. Lasa, J. Vergara, and I. Martínez de Marañón. 2008. "From the Chef's Mind to the Dish: How Scientific Approaches Facilitate the Creative Process." *Food Biophysics* 3, no. 2:261–268.

Auvray, M., and C. Spence. 2008. "The Multisensory Perception of Flavor." *Consciousness and Cognition* 17, no. 3:1016–1031.

Mintz, S., and C. M. Du Bois. 2002. "The Anthropology of Food and Eating." *Annual Review of Anthropology* 31:99–119.

Sacks, Oliver. 2004. "In the River of Consciousness." *New York Review of Books*, January 15, 41–44.

thirty-two

ON THE FALLACY OF COOKING FROM SCRATCH

CÉSAR VEGA AND DAVID J. MCCLEMENTS

while making a cake for my niece and nephew's birthday, I (Vega) realized that even though everybody thought I had baked the cake from "scratch," this was indeed far from the truth. Taking a closer look at the list of ingredients that made up the delicious clementine cake (eggs, sugar, cream cheese, butter, whipping cream, flour, corn starch, and, of course, clementines), it must be obvious to the reader that, beyond the eggs and the clementines, all the other ingredients—sugarcane, milk, wheat flour, and corn starch—are the result of a complex transformation of the original raw materials. But is it important or even relevant that we, as consumers, know this? What does it have to do with cooking from scratch, anyway?

The short answers to these questions are *yes* and *a lot*!

Food plays a central role in all our lives. An increased awareness of what we eat, where it comes from, and how it is transformed is critical in fostering a healthy and culturally rich society. The current debate around these issues, mainly among food scholars and activists, is certainly welcome, and we can recommend a few of the many good books published on the subject. But we also suggest complementing these books with information on the science of food and cooking, which can only enrich your culinary perspective. We say this because, as can be found in the multiple works of Michael Pollan—primarily, *The Omnivore's Dilemma*, *In Defense of Food*, and *Food Rules*—the terms *food science* and *processed foods* are used pejoratively. He goes so far as to suggest not buying foods with more than five ingredients or any ingredients you cannot easily pronounce. The statement was probably meant to make a point, but it could have critical implications if taken seriously. For example, enriching white flour with essential vitamins and minerals (such as niacin, folic acid, thiamine, iodine, calcium, and iron) has helped reduce the incidence of disease, such as pellagra, rickets, beriberi, and spinal tube defects. The removal of many of these essential additives from food products could have adverse effects on public health, particularly in economically challenged populations. There is incontrovertible evidence that food fortification with vitamins and minerals—which is based on a strong understanding of the scientific and health issues involved—has vastly improved the health of the population.

Nutrition(ism)

By contrast, there are an increasing number of food products that are enriched with *nonessential* nutrients. These are claimed to have health benefits beyond their basic nutritive value. Products like these are considered to be developed under the umbrella of what Pollan (2008) critically calls nutritionism: the belief that the nutritive value of food is simply the result of the sum of its parts. This, of course, negates the importance of the intrinsic structural and compositional complexity of food. A medium-size apple, for example, *does not equal* 5 ounces (150 g) of water + 11 g of fructose + 4 g sucrose + 4 g glucose + 4 g of dietary fiber + its share of vitamins and minerals, for a total of 95 calories. A number of other writers—Marion Nestle (2002) and Ben Goldacre (2008) included—agree with Pollan regarding the potential risks associated with nutritionism. One of the most important criticisms brought by these writers is that academic scientists, journalists, regulators, and the food industry frequently and irresponsibly extrapolate scientific results—obtained within the confines of very narrow experimental conditions—to dietary recommendations for humans in complex social settings.

We are of the opinion that the scientific evidence is simply not strong enough to make these claims for most of the nutrients studied thus far. The irresponsible social reach of such claims is then exacerbated by the myriad unsubstantiated articles appearing in newspapers and magazines championing the ability of some miracle food that can prevent cancer, slow down aging, or prevent or even cure certain diseases. As a result, consumers are overloaded with conflicting, confusing, and potentially misleading information, and they are finding it difficult to make properly informed dietary choices. On top of this, many sectors of the food industry are exploiting the increased public awareness of the potential health benefits of certain nonessential nutrients. Companies are doing this to differentiate themselves from their competitors and gain a marketing advantage. A health claim on a product label may give the impression that it is healthier (if it is healthy at all) than it actually is, thus encouraging consumers to eat more of it than would be consistent with a healthy diet. For example, if consumers rationalize an increased consumption of sugary soft drinks because they contain vitamins, this is likely to be detrimental to their overall health because of the high caloric density of these drinks. However, there may be benefits to consuming a fortified low-calorie soft drink rather than its high-calorie sugary alternative. Having acknowledged the limitations of nutritionism, it is important to highlight that fortifying foods with certain bioactive components may have health benefits for consumers, but more scientific evidence is still required before definitive health claims can be made.

There is not enough space here to go into the controversial topic of nutritionism at length. We refer the reader to the recent work of David J. McClements, César Vega, Anne E. McBride, and Eric Decker (2011), where a more thorough discussion surrounding the benefits of science in the design of healthier foods is provided.

Anyone Can Cook!

Michael Pollan (2008) proposes that all of us cultivate our own fruit, vegetable, and herb gardens as a means of increasing our intake of fresh and healthy foods. However, as Pollan himself acknowledges, not everyone has the will, resources, time, or finances to grow, prepare, and cook their own meals.

Let us emphasize the growing and acquiring part of the cooking equation for a moment. Let us assume that people are willing to and have the resources (time and money) to cook their own meals as well as the ability, because, as Chef Gusteau in the film *Ratatouille* says, "Anyone can cook!" In this acquiring and cooking exercise, the menu will be French onion soup, Greek-style salad, lasagna, and clementine cake. For the sake of discussion and argument, let us suppose that we have a whole vegetable garden in our backyard so that we do not need to go buy our vegetables. For a party of four, we need about 3 to 4 pounds (1.4–1.8 kg) of onions and a volunteer to slice them. Butter or oil is needed to slowly caramelize the onions (where the more correct term would be *Maillardize* as opposed to *caramelize*; see chapter 13). Unless you have a cow in your backyard and a centrifugal separator or churning machine, you might as well write down butter on your grocery list. The same applies for the oil: unless you happen to own a field of soybeans or canola—whose seeds are pressed to squeeze out the oil that would then be filtered, deodorized to remove unpleasant odors, and packaged—you will need to buy the oil. As for the eggs for the pasta and cake, Suzie and Dora, our laying hens, will provide. But what about the flour? Baking powder? Pasta sauce?

We hope that this blunt and brief example has at least made you reflect on the following:

1. Considering ourselves true omnivores as Pollan does, it is practically impossible (unless you are farmer, and even so) to adopt his eating style.
2. Cooking from scratch is a fallacy, unless we redefine what scratch really means.
3. Today, life without processed food is not possible.
4. Not everybody can or wants to cook!

Food in Context

Food scholar Rachel Laudan (2001) emphatically says: "No one with a 'metate' [mortar] will ever produce chocolate as smooth as that produced by conching in a machine for 5–10 hours." Let us emphasize that conching is only one of the many steps cacao goes through to be transformed into chocolate. We are unable to make chocolate at home, as well as a large number of other everyday foods. Consequently, there is still a need for convenient and inexpensive mass-produced foods for a majority of the general population. We are not saying that a diet should be based solely on processed food—wholesome foods, such as fruits and vegetables, should be part of any healthy diet. However, be aware that among these, quality can vary tremendously, especially from season to season. We believe that it is more important to distinguish good-quality versus poor-quality processed foods, rather than focusing on processed versus whole food as Pollan does. As exemplified earlier, many of the foods that we consider natural or whole are in fact processed in one way or another. This processing can include a variety of different physical, chemical, and biological actions, including mixing, blending, baking, frying, grilling, chilling, coating, adjusting of pH, adding salt, coloring, and fermenting. Therefore, we should not label all processed foods bad and all whole foods good. Instead, we should endeavor to understand how specific forms of processing alter the quality and nutritional properties of particular foods. A *potential* way forward might be found by categorizing foods based on their calorie-for-nutrient (CFN) quality, by which we could then rank their *relative* healthiness. The CFN is an index of value of thirteen key nutrients. Based on this argument, a food with a low CFN—for example, lettuce (CFN = 1.33)—would be of slightly better quality overall than the same amount of tomatoes (CFN = 4.12), better than skim milk (CFN = 8.12), and much better than white bread (CFN = 22.82). However, bear in mind that healthiness of food does not necessarily translate into healthy diets—reducing our food choices to any single scoring system could result into dietary imbalances.

A case can always be made that many processed foods are nutritionally poor and that overconsumption of these foods can contribute to unbalanced diets and their inherent health risks. However, it is often ignored that the roots of most of this behavior—convenience-driven overconsumption—can be found in the prevalent mix of social, cultural, and economic circumstances the United States has found itself in during the past one hundred years. It would be naive to deny that the food industry shares part of the responsibility for the current obesity crisis, but it would be just as naive to ignore that it plays and will continue to play a big part in the solution. The work of food scientists, technologists, and nutritionists has greatly contributed to the wide variety of inexpensive, convenient, safe, and nutritious foods available in the marketplace. This work has expanded the diversity of foods available to consumers, while decreasing the time and effort required to prepare them. Advances in food science and technology have allowed people to purchase a diverse range of food products all year round that enable rich and balanced diets; it is a luxury that was once available only to the elite members of society. Indeed, the increased life spans of individuals in Western countries over the past century can be attributed to better nutrition from the wider availability of inexpensive processed foods.

Better Food Through Science

We strongly believe that the proper use of the scientific method in the areas of food design, production, and distribution can be of great benefit to society. Indeed, as candidly put by C. P. Snow in *The Two Cultures and the Scientific Revolution* (1959) and extrapolated to the realm of food, this is our responsibility as scientists. The benefits—which are often overlooked or taken for granted—range from the production of a diverse range of ingredients and foods to the generation of stimulating insights into the reasons different foods look, taste, and feel the way they do. This knowledge complements the insightful observations of food scholars around *why* we eat *what* we eat.

As scientists, with a deep sense of responsibility toward the community we live in, we strongly oppose Pollan's (2008) and other food writers' and activists' denigration of the food science profession. One would get the impression that food scientists spend all their time in corporate laboratories creating "foodlike substances" to trick consumers into purchasing more fat, sugar, and salt. This is far from fair and does not give an accurate and thorough view of what food scientists actually do. It is true that some food scientists work for food companies, developing or improving processed foods that are high in fat, sugar, and salt. It is also true that overconsumption of these foods leads to a poor overall diet that negatively impacts health. Nevertheless, even these foods—for example, ice cream, potato chips (crisps), soda, and hamburgers—can be enjoyed for their desirable sensory attributes if they are consumed in moderation. Food scientists are involved in many other activities that demand the application of the basic principles of physics, chemistry, biology, and engineering to improve the manufacturing, storage, distribution, quality, safety, and nutritional attributes of foods.

Think, for example, about the multitude of naturally occurring chemical constituents within foods. And think about how these constituents interact with one another and with the human body to produce the characteristic physical and sensory attributes we associate with particular foods, including their unique taste, texture, and appearance but also their nutritional value. Food scientists work to understand this. Their aim is also to understand how foods and their constituents are changed by the various processes—mixing, kneading, stirring, frying, grilling, baking, boiling, microwaving, chilling, freezing—we subject them to during food preparation (at the factory and at home). The fundamental knowledge gained from these studies greatly benefits society, particularly those with limited access to food.

There are also many food scientists involved in understanding the complex physicochemical, physiological, and psychological processes underpinning the sensory perception of foods, such as appearance, texture, mouthfeel, and flavor. These scientists are trying to answer questions such as: What makes a food taste creamy, thick, or rich? Why does the flavor of a food change when the fat is reduced? What is the relationship between taste and appearance? Why are apples crisp or cookies crunchy? (Answers to some of these questions can be found in chapters 2 and 9.) Establishing the fundamental scientific basis for how foods are perceived will enable the design and manufacture of foods that are both of high quality and nutritionally responsible—for example, lower in fat, salt, and calories, while providing a pleasant sensory experience.

One of the most exciting new areas in food science today is a hybrid of sensory science, chemistry, and physics: molecular gastronomy. This food science amalgam came out of the realization that the phenomena occurring during cooking and eating were neglected by physical chemists. Today, its objectives are the understanding of the technical, artistic, and social components of cooking. In this sense, molecular gastronomy truly attempts to bridge the gap or, more correctly, close the loop among the sciences, humanities, and arts. It should also be noted that it is not a type of cooking, as is commonly and wrongly assumed (see chapter 30). Answering questions such as why ouzo (pastis) goes white when water is added to it, why the pulp around tomato seeds is so tasty, and why it is better to whip egg whites in a copper bowl is the objective of molecular gastronomy. Food science enables simpler and more daring cooking by clarifying and removing old culinary mysteries, which in turn can invite more people like you and like us into the kitchen.

Where Do We Go from Here?

The changing role of women in the kitchen over the past century or so is a key factor that at least partially explains the changes in Americans' relationship to food. Traditionally, the planning, preparation, and cooking of food was an arduous process that could take up a significant part of a woman's day. Innovations in home economics and the introduction of processed foods helped reduce the time and effort required to prepare meals. This increased convenience has been particularly important to many working families who have little spare time because both parents have to work to sustain the household. It is ironic that processed food is often unfavorably contrasted to home-prepared food, given that home cooks use a wide diversity of ingredients in their kitchens that are the result of food processing. These ingredients include vinegars, flavorings, oils, milk, cream, butter, margarine, chocolate, sugar, molasses, soy sauce, baking powder, maple syrup, flour, and so on. Even fruits and vegetables have been picked, stored, transported, and sold under carefully controlled conditions optimized through scientific studies to maintain their freshness and quality prior to consumption. Within this context, it is important to stop and reflect that food, through the act of cooking, is undeniably attached to culture. Cooking has the power to reunite the human being with nature; it makes us appreciate nature's bounty and kindness. Preparing food can be an enjoyable, creative, and rewarding activity, and a good meal can be an important family occasion. However, priorities vary among us. A large proportion of the *American* population choose not to devote their attention to the eating experience—the conscious selection, preparation, consumption, and enjoyment of food. One consequence of this choice may be an unhealthy diet. It is therefore important to acknowledge responsibility at the personal level for our diets and health.

Imagine that all processed foods and beverages were suddenly removed from the shelves of your kitchen and supermarket. What would you eat and drink? Where would you get your food? How much would it cost you? Could you trust it to be safe and nutritious? Would the food be available in winter? Would there be enough food to feed you and all your neighbors? How much time would you be willing to spend gathering, storing, and preparing your food? Considering these questions highlights how the application of science and technology to the manufacture, storage, and transport of foods and beverages has undoubtedly had a positive impact on our lives.

While many of the problems exposed by Pollan (2008) are legitimate, their roots (and solutions) reside mainly within sociological, cultural, and personal contexts. There is nothing inherently wrong with a food being inexpensive or convenient, but sacrifices in quality, nutrition, animal welfare, and sustainability must often be made to achieve this. The question then is: What compromises among quality, nutrition, cost, and convenience are consumers and society willing to accept?

We see good science as an essential part of the solution to the problems associated with the modern food system, like improving the ways nutrition-related claims are reviewed, boosting food quality, addressing the sustainability issue, and reducing the heavy reliance on fossil fuels and water in food manufacturing. Improvements to the food system require that consumers demand higher-quality foods (and a willingness to invest in their health, through food), governments establish appropriate guidelines and regulations, and society fosters a culture of food (through radical changes in school-lunch programs, for example)—and that the food industry responds to meet these demands by using science wisely. There is no better way to put it: each and every one of us has the power to make this happen by choosing wisely and, hence, responsibly, three times a day, every day.

Further Reading

Belasco, Warren. 2008. *Food: The Key Concepts*. New York: Berg.

Goldacre, Ben. 2008. *Bad Science*. London: Fourth Estate.

Guthman, J. 2007. "Can't Stomach It: How Michael Pollan et al. Made Me Want to Eat Cheetos." *Gastronomica* 7, no. 3:75–79.

Laudan, R. 2001. "A Plea for Culinary Modernism: Why We Should Love New, Fast, Processed Food." *Gastronomica* 1, no. 1:36–44.

McClements, D. J., C. Vega, A. McBride, and E. A. Decker. 2011. "In Defense of Food Science." *Gastronomica* 11, no. 2:76–84.

Nestle, Marion. 2002. *Food Politics: How the Food Industry Influences Nutrition and Health*. Berkeley: University of California Press.

Pollan, Michael. 2006. *The Omnivore's Dilemma: A Natural History of Four Meals*. New York: Penguin Press.

——. 2008. *In Defense of Food: An Eater's Manifesto*. New York: Penguin Press.

——. 2009. *Food Rules: An Eater's Manual*. New York: Penguin Books.

Vileisis, Ann. 2008. *Kitchen Literacy: How We Lost Knowledge of Where Food Comes From and Why We Need to Get It Back*. Washington, D.C.: Island Press.

thirty-three

SCIENCE AND COOKING

Looking Beyond the Trends to Apply a Personal, Practical Approach

MICHAEL LAISKONIS

i have been cooking professionally in a variety of capacities for nearly fifteen years, as a baker, line cook, sous-chef, and pastry chef. Though I fell into the business quite by accident, my own passion for cooking developed just as the arc of "foodie" culture and the celebrity chef began to shoot upward. I never planned to make a career of it, but looking back, I cannot imagine doing anything else. And as I say to any young cook starting out, it is truly an exciting time to be a chef.

In the past decade and a half, we have experienced a burgeoning interest in cooking and eating—both at home and in restaurants. This is in large part because our knowledge of and accessibility to once-exotic ingredients have increased, as has the phenomenon of reality television programs that present cooking as a spectator sport. While network and cable channels, not to mention print magazines and countless blogs, are creating more and better amateur cooks at home, the trends in professional cooking are cycling exponentially faster the more exposure these trends receive. The Internet plays a significant part in this as well. In addition to being a tool for research and instant information, it has provided a new form of dialogue among chefs seeking inspiration. Where chefs once toiled in relative obscurity and isolation, they have become a visible part of the growing global terrain. This global connectedness and exchange of ideas has in no small way transformed modern gastronomy.

But these trends have a tendency to define, for better or worse, a chef and his or her "style" of cooking. Whether intentionally organized or superficially suggested by the food media, such trends group like-minded chefs and their ideas into what some might call movements. Just within the past decade, the culinary landscape has seen the fusion of Asian, Latin, and pan-Mediterranean techniques and ingredients. We have praised "slow" food and we have flirted with raw food. We have sought and sourced foodstuffs from all corners of the earth, both celebrating their unique sense of place and shifting their contexts, while also emphasizing the importance of seasonal and organic food produced in our own backyards. Also, without rejecting high-quality ingredients, we have seen chefs applying the technological additives of mass-produced food in inventive cooking. Whether at the tables of fine restaurants or in the aisles of our local supermarkets, we increasingly look at what we eat through the lenses of fashion, politics, health, ecology, and of course technology and economics.

Haute cuisine—that level of the art and craft of cooking at which many concepts live or die—has seen a recent dressing down and, at the same time, an increased emphasis on creativity and theater. Highly refined cooking is no longer at odds with a casual atmosphere, reflecting perhaps a shift in emphasis from the pomp and circumstance of fine dining to our ever more sophisticated palates. Yet high-end dining has evolved, too, as chefs delve deeper into conceptual cuisine (figure 62); not only is there a message or a story behind the composition of food, even a sense of humor or irony, but, as diners, we have become partners in what has long been a one-sided pursuit. We are given instructions on how to consume what we are fed, or at times we actively engage in some aspect of the cooking or service of our meal. Suddenly, a new dialogue has opened up, and the discovery at the table is just as important as the ideas being developed in the kitchen. With this new order of experimental cooking—the infusion of personality—comes a perceived homogeneous culture. Globalization implies, to some, a certain sameness; infusing personality into cooking creates an identity. Despite the best of intentions and the emphasis on sense of place, if we were to close our eyes while dining in most temples of gastronomy, we could be just about anywhere— Paris, New York, Barcelona, or Tokyo. Depending on your point of view, this could be positive or negative.

Figure 62 Ice cream and sorbet as part of an artfully presented dessert. (Photograph courtesy of Michael Harlan Turkell)

As with the other arts, the most exciting developments and innovations in cooking often invite controversy. Perhaps the most incendiary culinary movement of late is molecular gastronomy. The phrase was coined by Hervé This, following the workshops on physical and molecular gastronomy. These workshops, held in Erice, Sicily, were organized by Nicholas Kurti, Hervé This, and Elisabeth C. Thomas to promote the investigation of the chemical and physical aspects of cooking. Uneasy with the public perception of molecular gastronomy, most chefs have tried to distance themselves from that specific terminology; a more democratic label might be what others have dubbed science-based cooking. Whether perceived as a vague concept or precise methodology, the term—indeed, the very idea—is often met with apprehension, skepticism, or outright disdain. Yet, as other technologies evolve along with our lifestyles, it makes sense that our food—if not its makeup then at least its presentation—must evolve as well. Besides, so many ideas that seem classic to us today were at one time revolutionary: for example, mayonnaise, puff pastry, and ice cream. Everything that is old was once new. And rather than see the new approach as polarizing the chef community, one should best adopt certain elements of it. If our common goal simply remains creating the most delicious food we know how, we all have something to gain from such progress.

The slow but steady creep of science into cooking has exposed more than a few myths and established practices of yesteryear. But also, it has brought to the fine-dining table new ingredients and techniques that have stretched the bounds of creativity. In fact, many of the ideas I refer to as new are merely new to us chefs; the mass-market food industry we have mined recently has known many of these "secrets" for decades. In addition to their conventional ovens and refrigerators, chefs are playing with thermal circulators, centrifuges, and vacuum machines, all techniques that are routinely used in the food industry. Alongside grilling, poaching, and roasting, some kitchens are using techniques that involve enzymes, liquid nitrogen, and lasers, techniques associated with a laboratory environment rather than a kitchen. In this respect, a certain measure of fear is understandable when we consider adapting the work of laboratories into our kitchens.

Along with the adapted technology, a new aesthetic has evolved, and much of contemporary cooking addresses form as well as function. While there are surely countless culinary crimes committed in its name, we cannot necessarily lay blame on the core tenets of science-based cooking. The true value in its use is the very essence of science: pure knowledge. It is about the desire to better understand the natural world, which for chefs pertains to our ingredients and how we manipulate them. This approach to cooking ultimately serves to liberate the chef, providing us the language to best speak to the foods we work with, while also bestowing the tools and ability (perhaps most important, an informed responsibility) to best express (or in some cases, exploit) the inherent qualities these ingredients have to offer. But this is something food scientists have always known.

We have only to look at the role of the scientist to understand how beneficial his or her methods are when applied in the kitchen. The primary goal of the scientist is simply to observe the natural world. Recording what is observed gives us truth, or at the very least, a theory to build upon. From there, the scientist asks: What if? The question serves to create a hypothesis, followed by tests and experiments that prove or disprove it.

The question is whether this method should be applied to cooking: the transformation of raw materials into something with added value, be it taste, texture, nutrition, entertainment, or nostalgia. The scientist has an advantage in that he or she is armed with a conceptual framework in which to fit the properties and behavior under investigation. Rather than begin with a measurable degree of predictability, chefs must often rely on a cook-and-look approach in the hope that a happy accident will yield desirable results. As we apply the basic scientific method—which includes careful observation and measurement to formulate and test hypotheses, together with a fundamental knowledge of the ingredients themselves—cooking becomes more efficient and orderly. Of course, it must be noted that chefs are not scientists; there will always be an appreciable gulf separating the two different disciplines. Science can certainly provide answers to chefs, but perhaps its greatest gift is the ability to ask the right *questions*.

My early days of cooking in the mid-1990s happened to coincide with the early days of molecular gastronomy and science-based haute cuisine. I was just coming up through the kitchen ranks and vividly recall reading short but intriguing dispatches from the workshops on physical and molecular gastronomy. I began to recognize and file away the names of Hervé This, Harold McGee, Heston Blumenthal, Pierre Gagnaire, and others. Also, rumors of a food revolution were being whispered in connection with an off-the-beaten-path restaurant in Spain called El Bulli and its mad genius of a chef, Ferran Adrià. Ever since, I have been in a position to observe, from varying degrees of distance and with a clear awareness and interest, the growing reach of this movement. I began to study the chefs who were practicing elements of this new brand of cooking; though linked by philosophy and spirit, they were ultimately expressing their own ideas and unique personalities. It seemed as if, in an instant, the culinary horizon appeared infinite.

Though over the years I would continue to wet my feet in this pool of discourse, I would resist diving in headfirst, lest it come to define my own style or approach to cooking. It just always seemed like a luxury only certain chefs could afford to practice, an aesthetic or skill set that could not be applied to my own environment, or simply a level of creativity I had yet to reach. This style of cuisine needed greater context. It was less a meal and more about the destination, the experience. Yet I have come to heed the siren call—at least in response to how it sounds to my ears—that sings of pure knowledge and understanding. This is the story of how I stopped worrying about whether I was the most creative or avant-garde or up to date. This is about how I came to embrace science for the plain fact that it allowed me to become a better cook.

The infusion of science into my everyday work has been incremental. Also, my interest in peeling back these layers of knowledge has coincided with my own culinary maturity and evolution as a chef. Experience has given me the confidence to use new techniques and ingredients to their best effect. And when I use the word *science*, it must be noted that I am not exactly tinkering around with vials and test tubes. In a real working kitchen organized around the structure of classical cooking, a scientific approach can simply mean looking just beyond the obvious, questioning the accepted logic, and most important, asking questions such as Why? and What if? In a sense, it is just knowing how to fix something that is broken. But then it is also taking something that might not necessarily be broken and breaking it regardless to see what happens or putting it back together in a different and useful way. For me, science-based cooking is collecting as much data as possible in the hope that the raw information will lead to something, if not new, then at least (and perhaps most crucial) more refined—tastier, creamier, lighter, crunchier—or better yet, that it will lead to something more delicious. In other words, learning as much as possible about an ingredient or process helps me better understand how it works in a recipe, thus opening my mind to applying that fact in a new way.

In reality, I have been no stranger to the calculated methodology we associate with science. As a pastry chef, precise measurements and an awareness (if only superficial) of the myriad chemical and physical reactions at play have, to some extent, become part of my inner fiber. It is true that all cooks must feel and observe as they cook, but for a pastry chef, it can be far less immediate and visceral. The calculation of a cake formula is like trying to see into the future, to predict the light and fluffy outcome of a dense, thick batter. Though I have always worked on the periphery of molecular trends, I am certainly no stranger to the use of hydrocolloids, *sous vide* (low-temperature cooking in plastic vacuum-sealed bags), and the rest. But my tendency was always to view these techniques, ingredients, and ideas in isolation as a means to a very deliberate end. Even professional chefs fall prey to the minutiae a recipe provides. Rather than step back to understand the interactions at play, or the flexibility of the "basic ratio," chefs can become slaves to the codified order of the recipe, making them unable to identify or solve problems that arise. Working with basic ingredient ratios, as opposed to fixed recipes, allows for flexibility in the creation of new dishes. For "ratio" we understand a set of ingredient-to-ingredient proportions that ensures reproducibility in the making of certain dishes or food products. For example, a basic ratio of 5 parts flour to 3 parts water (plus salt and yeast) will yield a workable bread dough, which we then can refine or adapt. Without such a holistic perspective, the chef will never be able to adjust the recipe to account for a new environment or to meet a specific need.

Rather than latch on to modern cooking for the flashy fireworks, the sexy aesthetic, or an avant-garde sensibility, we need to stick with the basics: I began to realize there was still so much to learn about the most basic ingredients we work with every day. Rather than channel all of my energy toward chasing the fads, tracking down the latest starch or gum, or seeking constant manipulation, more important was the knowledge of how to refine what I was already doing. I wanted to better understand the cause and effect of cooking, and wrap my head around the fact that all our raw materials are really just complex dispersions of even more basic components. Most chefs just do not think to break things down further than the fish or carrot. And while pastry chefs are already considered highly specialized within the culinary strata, once we start to immerse ourselves in the complexities of specific products, it is not surprising that many of us streamline our work even further, dedicating our careers to bread, chocolate, sugar, or ice cream. Sometimes, narrowing our options is the only way to open up new opportunities.

Interestingly enough, the one ingredient that really got me thinking deeper was milk. It seems somehow appropriate to zero in on the most primal of foodstuffs—the one thing that unifies us humans as a species, no matter our location or culture. Granted, we are the only animal to continue consuming it (and milk taken from other beasts at that) long after the need for it is gone. Indeed, as a pastry chef interested in the physiology of taste and our craving for sweet, it would not surprise me if our lifelong penchant for sweet is somehow rooted in that very first taste of mother's milk. Yet we continue to rely on its nutrients and enjoy its subtle, comforting flavor. We exploit its physical and chemical properties in all manner of products, and we have discovered countless functions and methods for processing and preserving it. Its ubiquitous presence in the cooking of nearly every culture is perhaps equaled only by eggs and wheat—and no doubt rivaled only by salt. The complexity and variability of milk's seemingly simple constituents affect so many of the products we create that we have long taken the staple for granted. Milk was not only just a fascinating product I used every day in my work, but it served as a symbolic and fundamental launching pad for developing my own personal approach to cooking.

Technically, my investigation began not with milk but with one of its end products, butter. For several years, I had been making ice cream infused with brown butter. While it lends a nutty complexity to sauces and sautés, it might not be the first flavor one associates with pastry. Yet it is one of my favorites, and as a component in high-end desserts, one might argue that it is a bit trendy at the moment. The ice cream recipe that I had stumbled upon (and used without question) called for the addition of browned butter to a base of milk, sugar, and egg yolks, assisted by the addition of cornstarch and a stabilizer blend. Introducing so much flavorful fat made for a base that walked a textural tightrope; if overprocessed at the freezing stage, the ice cream could easily turn into an unpleasant, grainy mess, presumably due to phase separation. But I clung to the recipe nonetheless, defending the ice cream's amazing flavor while simply accepting its shortcomings. Until I fully grasped all the variables at work, there was little I could do to solve the problem. I was, in short, a slave to the recipe. The solution would eventually present itself on three fronts as I focused on the building blocks of ice cream, recalled a long-forgotten cooking technique, and took a look at the basic composition of each of my ingredients.

I had been a working pastry chef for two or three years before I ever *really* understood ice cream. I had certainly made ice creams and sorbets in the earliest days of my career, but looking back at some of those early formulas, it is surprising how out of whack the recipes were, at least in relation to how I construct ice cream now.

One of the most fundamental bases in pastry work is the custard. One might even suggest that crème anglaise should have been codified alongside Georges Auguste Escoffier's "mother sauces." We are initially introduced to the custard family by way of crème anglaise—scalded milk and cream tempered into egg yolks and sugar and then gently cooked to a lightly thickened consistency. Though the basic ratios might differ slightly, the rest follow suit: add starch and you have pastry cream, and when baked, that custard becomes crème brûlée, and so on. Conventional wisdom then told us if we freeze that initial crème anglaise base, voilà, we have ice cream. Many pastry chefs still follow that dictum, or if they use a recipe that fine-tunes the components, they do so without full knowledge of why they do it or how those ingredients function. And the one recipe might be interchangeable, even with the addition of chocolate or fruit or alcohol. Pastry chefs stuck with ice cream and sorbet recipes that worked and made adjustments based on taste—a classic example of the cook-and-look approach. Nothing wrong with that, I guess, unless you want some measure of consistency from batch to batch.

Rather than looking at ice cream as a recipe calling for milk, cream, sugar, and eggs, we are better served by looking at a formula composed of water, fats, and solids (sugars and proteins, plus added or naturally occurring stabilizers and emulsifiers); we are then, of course, forced to better understand the makeup of our ingredients—complex dispersions themselves—and what each of these basic components brings to the final product.

I am jealous of chefs starting out today. They now have greater access to several important books on the subject of ice cream; favorites of mine include Francisco Migoya's *Frozen Desserts* and Angelo Corvitto's *Los secretos de helado* (*The Secrets of Ice Cream*). This knowledge is not exactly new to the dairy industry, but before these easy-to-read references—and especially before the Internet—such technical information was difficult for chefs to find. Cooking schools are finally adopting a more formulaic approach as well. As I started to pay closer attention to sorbet and ice cream "technology," I noticed how these new recipes differed from mine. Though I did not yet understand *why* they were better, I saw the patterns: a higher proportion of milk to cream, added nonfat milk solids, the use of glucose and invert sugar syrup in addition to sucrose. Most important was that the result was better than anything I had produced previously. When prominent pastry chefs (Olivier Bajard, Oriol Balaguer, and others) began sharing their own research in books and seminars, it became clear why their ice creams were better; it was from these base recipes that I adapted a starting point of my own.

I slowly came to respect the intricacies and the science behind the ingredients we work with, and how knowing their properties can help us create recipes that are really more like mathematical equations. I first saw Olivier Bajard demonstrate such an equation with simple chocolate mousse—a dispersion of ingredients that in themselves are actually complex dispersions. Whether it is ice cream or mousse, once we accept a few general guidelines with respect to fat content, dry matter, and sweetening power, and then acquaint ourselves with the makeup of our ingredients, we immediately see how everything falls into a perfect equation. And once we have an equation, we have at our disposal an indispensable tool for adjusting each variable when a change is made to just one of them (for further explanations and examples of the use of an "ice cream equation," see chapter 17).

The challenging part of this process, but equally exciting, is what we discover by merely compiling all of this data on food composition (figure 63). My field of research suddenly expanded well beyond

cookbooks and into the realm of nutrition charts, research papers, biochemistry textbooks, and patent applications. Before tackling the brown butter ice cream in particular, I began reengineering all of my old recipes. I applied the basic rules, all of which combine to create optimum texture and stability. Entire books have been written describing all the intricate calculations and functionalities of ice cream's constituents, but with just a few ratios and percentages I suddenly had the ability to exercise some element of control over this important aspect of my craft.

Ideally, the fat content of a pleasantly textured ice cream should lie between 7 and 12 percent of its entire yield. Given a simple recipe to evaluate, how do

Figure 63 Using a calculator and data on dairy composition to build an ice cream formula.

I determine its fat content? Well, knowing that whole milk contains slightly less than 4 percent fat and heavy cream 36 percent, I can determine the butyric fat (fat from milk products) by simply adding the respective percentages of the weight of those ingredients. Then, knowing that fat makes up one-third of an egg yolk, I can add in the weight of its fat. Now on to the dry matter, which is crucial in determining the overall texture of the finished ice cream. Dry matter is everything in the mix that is not water; the nonfat solids, most of which are sugar, depress the freezing point of the water in the mix by locking it away, rendering it unable to form ice crystals. These nonfat solids, plus the fat content determined previously, typically make up an ideal percentage in the range of 37 to 42.

When we look ever closer at ingredient composition, we realize that dry milk powder actually contains a small amount of water, as do semiliquid ingredients like glucose and invert sugar (generally about 75 percent solid material and 25 percent water). Even our liquid ingredients—milk, cream, and egg yolks—are made up of nonfat solids that must also be figured into our target percentage. Once we know that about 8 percent of the milk, 6 percent of the heavy cream, and 20 percent of the quantity of egg yolks in the recipe are made up of nonfat dry matter, we can accurately arrive at our final composition.

And there are other considerations: balancing the sweetening power with the solid content of the sugars, determining the amount of added nonfat milk solids, and of course, figuring the correct dosage of stabilizer (helpful in preventing unpleasantly large ice crystals). Every substitution or variable added down the road will require adjustments across the board. Further, we might fine-tune a formula if the method of production or the equipment changes. I typically use a conventional batch freezer, in which the mix is rapidly frozen while spun (figure 64). More recent technology, like the Pacojet, can

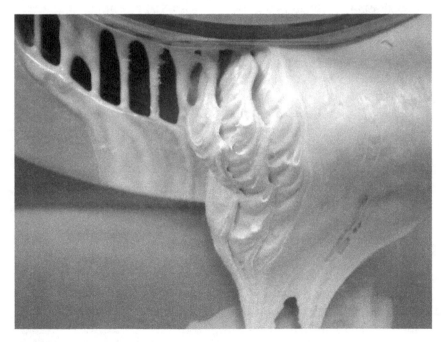

Figure 64 Extraction of ice cream from a restaurant batch freezer.

produce ice cream in a very different way. In essence a pressurized, high-powered food processor, the Pacojet works in reverse, milling or shaving a prefrozen base. This was my initial eureka moment—witnessing how crucial the knowledge of ingredients and the processes involved is, and how it all certainly comes under the umbrella of science-based cooking.

The original recipe I had been using for the brown butter ice cream accounted for some of this delicate balance—there is a fair amount of pure butterfat balanced by lowfat milk and some added cornstarch (which was assumed to stabilize the emulsion)—but it ultimately fell short. A few ill-informed attempts to adjust the recipe still did not yield favorable results. However, now armed with the data and the knowledge to calculate the content of the solids and fat in dairy, it seemed logical to focus on the milk solids that make the flavor of the ice cream so special. So what is brown butter? It is butter—and nothing else—that is allowed to brown, right? But let us take a step back to ask ourselves, What is butter? Fat, of course. And in good-quality butter, fat makes up at least 82 percent of its weight. But fat alone does not brown; sure, you can burn a fat until it smokes and blackens, but that is a different kind of chemistry at work, and burned butter does not taste all that good! So, with 82 percent fat, that leaves 18 percent, and the water in butter accounts for most of it, about 16 percent of the total. Nearly all that water cooks off or evaporates in the process of browning. So really, the most important constituent of butter is everything that is left: the measly 2 percent that is composed of milk solids. When clarifying butter, these proteins and sugars are the scum and foam that we carefully skim off to get at the pure butterfat. But here we want to leave them intact, and we want them to brown to create the trademark flavor we seek.

How does that happen? Once that 16 percent (the water) has cooked off into the atmosphere, we are left with just the fat and the solids. Remember, at least at sea level, liquid water boils and turns to steam at 212°F (100°C). Once that water is out, the fat can finally exceed that temperature barrier. As it begins to rise, about another 100°F (38°C) or so, we begin to see chemical reactions take place. These reactions are known as the Maillard reaction, named for Louis-Camille Maillard, the scientist who discovered them. In simple terms, these reactions are responsible for the phenomenon of color and flavor formation resulting from the effect of heat on mixtures of proteins and sugars (for more on the Maillard reaction, see chapter 13). It is the Maillard reaction that makes a perfectly cooked steak taste more appealing and complex than a piece of raw meat. But back to our butter. The nonfat solids, for lack of a better term, fry in the butterfat, and the fat in turn becomes infused with the flavors of the browned solids. However, we need to remember that those flavor-generating solids make up only 2 percent of the butter's weight. As we start to consider other uses for our brown butter, we are confronted with the fact that those browned bits are dispersed in a lot of fat. To increase the intensity of flavor, it stands to reason that we would have to add more fat as well. This is especially challenging if we want to make, say, a brown butter ice cream or a brown butter ganache for chocolate bonbons. As we have seen with ice cream, these are applications in which fat has to be delicately balanced for both texture and flavor. So one step forward, two steps back: Where does butter come from? Milk of course, but more specifically, we might say cream.

We can make our own butter by overwhipping cream to the point where the fat globules jam themselves together, squeezing out a good deal of the water. And in that water—the original buttermilk—remains the majority of the milk solids. So if we were to compare equal measurements of butter and cream, we would discover that the cream contains 6 percent milk solids by weight, or three times more milk solids than those obtainable from butter. Not only can we extract a higher yield of milk solids to flavor our ice cream by using cream, but using cream is also far less expensive. As a side note, we could consider whole and skim milk, which contain even more milk solids, up to 8 or 9 percent. But milk contains a lot more water, too—water that would take a long time to cook off, unless of course we were working on an industrial scale. Indeed, in commercial evaporated and condensed milk, the solids content jumps to between 20 and 30 percent of the total weight. And then that leads us to dulce de leche, and why it tastes so good: it is all about the browned milk solids, the Maillard reaction.

So how do we go about browning and extracting the solids in heavy cream? It is the same process as for the butter, but as one would imagine, it takes longer. As more and more water evaporates, the natural emulsion of the cream breaks up, causing the nonfat solids to separate from the fat. These nonfat solids tend to clump together—there are three times more of them in cream than in butter—which, when carefully strained, also gives us a usable by-product of pure clarified butter: a sort of brown-butter "powder" that can be added to just about anything. There remains some amount of fat among the solids, but little enough to easily adjust for in our recipe. There is, however, a very simple alternative solution to increase the nonfat solids, one that I initially overlooked: one can simply increase the solids by adding extra nonfat milk powder. This eliminates having to stand over a pot of reducing cream or butter, and it also greatly boosts the yield of the nonfat solids.

Applying this sliver of dairy science was so exciting that I naturally wanted to look at other products and use this knowledge to expand my repertoire even further. Chocolate is another excellent case study because, in addition to being eaten as is, it adds value to many other products. But only with an understanding of its composition and microstructure can we best incorporate it into cakes, creams, and candies. In its most common form, more than half the weight of dark chocolate is made up of the cacao bean itself, which brings to the chocolate bar both the flavorful cocoa solids and fat in the form of cocoa butter. Sugar, of course, and usually small amounts of vanilla and lecithin make up the remainder. Milk chocolate will no doubt contain fewer cocoa solids, but they are replaced with more sugar, and of course, milk solids. On to white chocolate, which contains no cocoa solids apart from the cocoa butter, hence its off-white color and neutral flavor. But we must also include the sugar and milk solids, which, respectively, make up around 40 and 20 percent of its total weight. Consulting data on the composition of dairy products, we notice that white chocolate is remarkably similar to sweetened condensed milk, at least in terms of those two components, sugar (42 percent) and milk solids (23 percent). While few of us would get excited over the prospect of munching a block of white chocolate, exploiting its composition just might provide surprising results.

I am sure that most professional pastry chefs and home cooks have at least once accidentally scorched or ruined chocolate by overheating it, as I have done. By applying what we have learned from brown butter and reduced cream, we can take that accident and turn it into something delicious. Remember, white chocolate is composed of nothing more than cocoa butter, sugar, and milk solids, along with small amounts of vanilla and lecithin. Overheating white chocolate will lead to at least two types of reactions: one is the caramelization of the sugar in the white chocolate (figure 65), and the other is the Maillard reaction that occurs when the sugars and milk proteins in the nonfat milk solids interact together. Most chefs would discard the mistake of burned chocolate and start over again. It was not until recently that I realized there was an idea that could come from this.

Figure 65 A pistachio mousse impregnated with caramelized white chocolate. (Photograph courtesy of Michael Harlan Turkell)

I had attended a seminar led by the development team at the gourmet chocolate company Valrhona. The team presented its own intentional, controlled process of browning the sugars and milk solids in white chocolate—gently "roasting" them in a low oven to produce a complex, caramel flavor. Remember the similarities between the chocolate and the condensed milk? It all makes sense when we consider that gently cooking condensed milk produces the same intense flavor profile. Such a transformation gives new life and lends sophistication to an otherwise flat and uninteresting ingredient. And of course, it is exciting to realize that others are working on the same problems in a different setting. In the case of caramelized chocolate, one of the world's largest chocolate producers sneaks the same technique, albeit in an industrial setting, into a handful of its mass-market products.

The best thing about being a chef is that I never exhaust my capacity for learning. Indeed, chefs are students for life, at least as long as they remain receptive to new ideas, ingredients, and techniques. This fresh and inquisitive attitude toward cooking makes me more passionate about the work I do, and with each new discovery, the foundation is laid for the next innovation, and the next, and so on. And taking part in the process necessary to attain this new knowledge is inspirational. In fact, being involved in the research, experimentation, and methodology that inspires the creative process beyond any singular result is doubly rewarding. Science continues to ask questions, and thus society continues to progress. I hope this same spirit and quest for truth finds increasing favor in the culinary community so that we, too, evolve. In the end, we will eat better for it.

Further Reading

Corvitto, Angelo. 2004. *Los secretos de helado*. Barcelona: Basch. Available at http://www.angelocorvitto.com.

Goff, Douglas. *Dairy Science and Technology*. Available at http://www.foodsci.uoguelph.ca/dairyedu/home.html.

McGee, Harold. 2008. "Modern Cooking and the Erice Workshops on Molecular and Physical Gastronomy." *Curious Cook*. Available at http://curiouscook.com/cook/erice.php.

Migoya, Francisco J. 2008. *Frozen Desserts*. Hoboken, N.J.: Wiley.

Pollan, Michael. 2008. *In Defense of Food: An Eater's Manifesto*. New York: Penguin Press.

Ruhlman, Michael. 2009. *Ratio: The Simple Codes Behind the Craft of Everyday Cooking*. New York: Scribner.

This, H. 2009. "Molecular Gastronomy: A Scientific Look at Cooking." *Accounts of Chemical Research* 42, no. 5:575–583.

Vega, C., and J. Ubbink. 2008. "Molecular Gastronomy: A Food Fad or Science Supporting Innovative Cuisine?" *Trends in Food Science and Technology* 19:372–382.

CONTRIBUTORS

ANDONI LUIS ADURIZ is the owner and chef of Mugaritz, a restaurant near San Sebastián, Spain, whose team includes Daniel Lasa, Oswaldo Oliva, and Javier Vergara. It is considered the third-best restaurant in the world, according to *Restaurant Magazine*. In 2005, *Michelin Guide* awarded Mugaritz its second star. In 2007, Harold McGee presented his latest book, *On Food and Cooking*, at Mugaritz. Considered one of the bibles in gastronomy, its preface was written by Aduriz.

JUAN-CARLOS ARBOLEYA obtained his doctorate in physical biochemistry at the University of East Anglia and started his scientific career at the Institute of Food Research, both in the United Kingdom. His current research at AZTI-Tecnalia Food Research Institute, in Spain, focuses on the fundamental mechanisms that control food texture to develop strategies for improving its sensory and nutritional properties. His expertise finds application in the design of new dishes in haute cuisine by means of collaborations with prestigious restaurants, such as Mugaritz near San Sebastián, Spain.

RAJEEV BHAT is working as a faculty member in the food technology division, School of Industrial Technology, at Universiti Sains Malaysia. He specializes in the area of food safety and food nutrition. Bhat has published more than seventy-five research papers in peer-reviewed international and national journals. He has written several book chapters and has coedited two books, on biotechnology and on food preservation technology.

JONNIE BOER is chef of De Librije, a Michelin three-star restaurant housed in the library of a former fifteenth-century abbey in Zwolle, the Netherlands.

MICHAEL BOM FRØST is an associate professor at the Sensory Science Group at the University of Copenhagen and directs the Master of Science program in gastronomy and health. Since 2007, Bom Frøst has worked in the area of molecular gastronomy. He is a well-recognized all-round sensory scientist with solid expertise in research that combines basic science with industrially relevant applications. He is keenly interested in the dissemination of sensory science and its relevance and application to other research, industry, and the general public.

ADAM BURBIDGE is group leader at the Nestlé Research Center in Lausanne, Switzerland, and visiting professor at the University of Swansea in the United Kingdom. He has a background in chemical engineering and food technology. One of Burbidge's passions is discussing science and food over a good dinner.

JULIA CALVARRO works in the Food Science Department at the University of Extremadura in Spain. She has an undergraduate degree in food science and a master's in meat science. As a doctoral candidate, she is studying nitrosamines and the flavor of meat products. Her master's thesis focused on the use of transglutaminase in gelatin gels and foams.

PERE CASTELLS obtained his degree in organic chemistry from the University of Barcelona. After teaching high school chemistry for twenty years (during which time he wrote several reference books), he started a food research collaboration with El Bulli restaurant in Roses, Spain. In 2004, he headed up the science and gastronomy division at ALICIA, a Spanish food research institute. Castells wrote, along with the Adrià brothers, the book *Modern Gastronomy A–Z* (2010), which has been translated into six languages. He is a scientific coordinator of the general education course Science & Cooking at Harvard University.

CRISTINA DE LORENZO is the head of the Department for Science and Technology Transfer at IMIDRA, an independent body whose main objective is research and innovation in the agriculture, livestock, and food industries. She received her degree in agronomy from the Polytechnic University of Madrid. She has eighteen years of experience in food-quality research, mainly in traditional regional foodstuffs from the Madrid region, via the use of biochemical, sensorial, microscopic, and molecular techniques.

SUSANA FISZMAN is a research professor at the Institute of Agrochemistry and Food Technology, IATA-CSIC, Spain. Her area of expertise is the physical properties of food. She has principally focused her research on texture studies—specifically, thickeners and new ingredients, analyzing their interaction in a number of food matrices and their relationship to sensory properties. She has published more than one hundred scientific papers and has written twelve chapters in specialized books. She is on the editorial board of the journal *Food Hydrocolloids*.

TIM J. FOSTER is associate professor and reader in food structure at the University of Nottingham in the United Kingdom. As a senior scientist, he led teams in new technology development and controlled delivery of flavor and health actives at Unilever for sixteen years. Foster aims to provide design rules for ingredient interchangeability and improvement and novel functionalities through the understanding of the physical properties of single and mixed polysaccharide systems—from polysaccharides in the plant cell wall to the influence of processing on structure–functionality in microstructure creation. His current focus is on natural structuring agents, rehydration phenomena, and food microstructure changes during digestion.

AMELIA FRAZIER works as a research-and-development scientist at General Mills. She earned a culinary associates degree from Kendall College in 1999 and a pastry certificate from the French Pastry School in 2001, both in Chicago. She worked as a pastry cook in several fine-dining establishments and as pastry chef at Frontera Grill and Topolobampo in Chicago. She graduated from the University of Wisconsin with bachelor's (2008) and master's (2011) degrees in food science.

MARÍA J. GÁLVEZ-RUIZ is a professor in the Applied Physics Department at the University of Granada and head of the Biocolloid and Fluid Physics Group. She holds an undergraduate degree in chemistry and a doctorate in physics from the University of Granada. Her focus is on the science and technology of colloids and interfaces and their application to the areas of food, pharmacy, and biomedicine.

MATT GOLDING is associate professor in food structure at Massey University in New Zealand.

JAN GROENEWOLD is assistant professor of physical chemistry at Utrecht University in the Netherlands. He combines this academic work with consulting for industry (www.denk-werk.nl). Groenewold and Eke Mariën are working in the field of molecular gastronomy. Lecturing and book writing have been their main activity under the flag "Cook & Chemist" (www.cookandchemist .com).

RICHARD HARTEL has been a professor at the University of Wisconsin–Madison for twenty-five years, teaching courses on food engineering, manufacturing and processing, and candy science. His research interests revolve around phase transitions in foods, including ice formation and recrystallization, sugar crystallization and glass transition, and lipid crystallization. Tracking down the many causes of bloom formation and inhibition in chocolates has been an ongoing area of research.

HELEN HOFSTEDE was a master's student in food physics at Wageningen University, the Netherlands. She then took an internship at the restaurant El Bulli, in Spain, and is now employed by Pepsico in the Netherlands.

ARIELLE JOHNSON is a doctoral student in agricultural and environmental chemistry at the University of California, Davis, where her research focuses on the chemical and sensory analysis of food and wine flavor. She became involved in the Experimental Cuisine Collective (ECC) as an undergraduate in chemistry at New York University and conducted an independent study on dondurma that led to her contribution to this book.

AKI KAMOZAWA and H. ALEXANDER TALBOT are the brains behind Ideas in Food, which is a blog, a book, and a culinary consulting business based in

Levittown, Pennsylvania. They specialize in sharing techniques that use modern ingredients, equipment, and innovative approaches in the preparation of food. In addition to their work with individual chefs and restaurants, they have consulted with organizations such as the Institute of Culinary Education in New York, Marks & Spencer, and Unilever. In 2010, Kamozawa and Talbot published their book *Ideas in Food: Great Recipes and Why They Work*. They are featured in the books *Modernist Cuisine* (2011) by Nathan Myhrvold, Chris Young, and Maxime Billet and *Cooking for Geeks* (2010) by Jeff Potter. They contributed to the anthology *Food and Philosophy* (2007) and wrote the column "Kitchen Alchemy" for Popular Science online in 2008.

ALIAS A. KARIM is a professor of food technology in the food technology division, School of Industrial Technology, at Universiti Sains Malaysia. He joined the university in 1994 and since then has taught most of the food science subjects in the curriculum. His research focus is mainly on the fundamental and applied aspects of structure–property relationships and technological applications of starch and nonstarch polysaccharides. He has published more than seventy papers in international peer-reviewed citation-indexed journals.

JENNIFER KIMMEL is a protein chemist at Kraft Foods, where she applies fundamental protein chemistry to support the company's cheese and dairy portfolio of products. She received her doctorate in biochemistry, specializing in enzyme regulation, from Texas A&M University. Prior to joining Kraft Foods in 2005, she spent four years researching enzyme mechanisms as a postdoctoral fellow at the University of Missouri–Columbia. Away from work, she is an avid CrossFitter and enjoys hiking at national parks.

KENT KIRSHENBAUM is associate professor of chemistry at New York University. His research interests include bioorganic chemistry, biomimetic chemistry, protein conformation and dynamics, macromolecular design—and food. He is one of the founders of the Experimental Cuisine Collective and regularly speaks to academic, popular, and media audiences about the chemistry of cooking.

TIMOTHY KNIGHT is employed at the Oscar Mayer division of Kraft Foods in Madison, Wisconsin, where he is a technical expert. He earned his doctorate, Master of Science, and Bachelor of Science degrees, all in food science and technology, from Texas A&M University, where he taught courses in food chemistry and instrumental analysis in addition to conducting fundamental research on many different commodity groups. His areas of expertise include food chemistry, analytical measures, meat science, dairy science, and food microbiology. His focus is on developing innovative process technologies, exploring crossover technologies from other fields, and gaining a fundamental understanding of ingredient- and process-based influences on food systems.

SERGIO LAGUARDA is part of the teaching staff at Escuela de Hostelería de Alcalá de Henares in Madrid, where he specializes in hazard analysis and the critical control points of food and takes part in Spanish pastry contests. He obtained a culinary degree from the High School of Cooking of Madrid.

He possesses extensive experience working in restaurants and hotels around Spain and London. His area of specialization is baked goods and pastry.

MICHAEL LAISKONIS, executive pastry chef of New York's Le Bernardin, was named *Bon Appétit*'s Pastry Chef of the Year in 2004 and Outstanding Pastry Chef in 2007 by the James Beard Foundation. His work has helped the restaurant maintain four stars from the *New York Times* and three stars from the esteemed *Michelin Guide*. In 2008, Laiskonis became a featured contributor to *Gourmet* and Salon. Most recently, his writing has appeared on the Huffington Post, in the *Atlantic*, and on two Web sites documenting his work: mlaiskonis.com and michael-laiskonis.com.

MARTIN LERSCH earned his doctorate in organometallic chemistry in 2006 at the University of Oslo. Since then, he has been working in research and development at the Borregaard biorefinery in Norway. Aside from his daytime job, he has a strong interest in molecular gastronomy and popular food science, and he has blogged about this at Khymos (http://blog.khymos.org) since 2007. During the past ten years, he has given numerous food- and kitchen-related popular science talks and has made several appearances in Norwegian newspapers and on national radio and television. Lersch can be contacted at webmaster@khymos.org.

JULIA MALDONADO-VALDERRAMA is a lecturer at the University of Granada. She earned her doctorate in physics from the University of Granada in 2006. She specialized in food nanoscience from 2006 to 2010 as a Marie Curie Fellow at the Institute of Food Research.

EKE MARIËN has been active as a chef and caterer. Currently, he is exploiting the food store Boerenjongens, specializing in locally produced foods. Jan Groenewold and Mariën are working in the field of molecular gastronomy. Lecturing and book writing have been their main activity under the flag "Cook & Chemist" (www.cookandchemist.com).

ANNE E. MCBRIDE is director of the Experimental Cuisine Collective (ECC), a group with more than seventeen hundred members that examines the relationship between food and science. Also, she is a doctoral student at New York University, where she researches the evolution and mediatization of professional cooking in the United States.

DAVID J. MCCLEMENTS is a professor in the Department of Food Science at the University of Massachusetts. He specializes in the areas of food biopolymers and colloids, particularly the development of food-based delivery systems for active components. McClements received his doctorate in food science at the University of Leeds, United Kingdom. He then conducted postdoctoral research at the University of Leeds; University of California, Davis; and University College Cork in Ireland. He has published or edited more than eight books and more than 450 scientific articles. He has received awards in recognition of his scientific achievements from the American Chemical Society, Institute of Food Technologists, American Oil Chemists Society, and University of Massachusetts.

LINE HOLLER MIELBY is a doctoral student in sensory science at Aarhus University in Denmark. Prior to that, she worked as a research assistant in the field of molecular gastronomy and the enjoyment of food. Mielby holds a master's in food science.

JOHN R. MITCHELL is professor emeritus of food technology at the University of Nottingham. Before joining the university, he worked for Unilever and Mars, Inc. He has supervised more than sixty doctoral students and published approximately two hundred papers. He received the Food Hydrocolloids Trust medal for his work on hydrocolloids in food and is an elected member of the Academy of the International Union of Food Science and Technology. He continues to be involved in research partly through the company Biopolymer Solutions, and he edits the international journal *Carbohydrate Polymers*.

LOUISE M. MORTENSEN is currently seeking a doctoral degree in molecular gastronomy in the Department of Food Science at the University of Copenhagen. The focus of her thesis is the low-temperature cooking of meat—specifically, studying the relationship between preparation time and temperature, and the resultant sensory characteristics. She has a master's degree in food science and technology from the University of Copenhagen. She has been employed in the manufacture of analytical instruments. Mortensen has cowritten a review paper on molecular gastronomy, which was published in *Chemical Reviews* in 2010. In addition, she is actively involved in establishing teaching in gastronomy at the University of Copenhagen.

NATHAN MYHRVOLD wrote, with Chris Young and Maxime Billet, *Modernist Cuisine* (2011). An accomplished cook, he is also CEO of Intellectual Ventures, a firm dedicated to creating and investing in invention. Before founding Intellectual Ventures, Myhrvold was the first chief technology officer at Microsoft. Myhrvold's formal education includes undergraduate degrees in mathematics, geophysics, and space physics; doctorates in mathematical economics and theoretical physics; and postdoctoral work in quantum theories of gravity with renowned cosmologist Stephen Hawking.

MIRIAM PETERS works as a process developer at FrieslandCampina, the Netherlands. She earned a bachelor's degree in food science from the University of Applied Sciences in Den Bosch, the Netherlands. She continued her studies at Wageningen University, receiving a Master of Science in product functionality.

MALCOLM POVEY is president of the University and College Union at Leeds University, United Kingdom, and one of the inventors of the Ultracane sonic mobility aid for the blind and the Baker Petrolite aggregation monitor. His group has recently gained prominence (via the Discovery channel, Sky News, and the BBC) from its discovery of a relationship between acoustic emission, crack propagation, and the human perception of crispness. Povey's laboratory uses the most advanced techniques for the characterization of nanoparticles and colloidal particles, including ultrasound microscopy and spatial scanning (the Acoustiscan). Current publications can be found at www.food.leeds.ac.uk/mp.htm.

CHRISTOS RITZOULIS has held the position of senior lecturer in food chemistry at the Department of Food Technology of ATEI in Thessaloniki, Greece, since 2008. He obtained his undergraduate degree in chemistry in 1996 at the Aristotle University of Thessaloniki, Greece, and his master's and doctorate in food science at the University of Leeds, United Kingdom, in 1997 and 2001, respectively. Ritzoulis has served as a Sergeant (chemist) for the Hellenic Army from 2001 to 2003, subsequently working as a postdoctoral researcher at Aristotle University for a year. From 2004 to 2008, he worked as an analyst for the Hellenic States General Chemical Laboratory.

JORGE RUIZ earned his doctorate in the study of Iberian ham flavor. He was a postdoctoral student at the Foulum Research Center in Denmark and at Cornell University in the United States. This postdoctoral work focused on the use of solid-phase microextraction (SPME) for analysis of meat volatiles and the investigation of transglutaminase as a means for improving the functionality of meat proteins. In 1999, he joined one of the molecular gastronomy meetings at Erice, Sicily, and since then, part of his research-and-development activities has been devoted to culinary applications. He is the author of numerous scientific papers and chapters and currently chief researcher of a Spanish science and gastronomy network of researchers.

NATALIE RUSS, a student of chemistry at the University of Mainz, Germany, entered the food laboratory at the Max Planck Institute for Polymer Research in 2007. She uses thermal analysis, rheology, flow dynamics, and mechanical experiments as main techniques. She has been very much involved in helping to develop the food laboratory.

ELKE SCHOLTEN works as an assistant professor in the Food Physics Group at Wageningen University, the Netherlands. She has a Master of Science in physical chemistry from Utrecht University, also in the Netherlands, after which she completed her doctorate at Wageningen University in food technology. One of her interests is to understand physical phenomena in food products with respect to gastronomy and how to use this knowledge to manipulate texture and sensory properties.

SIDNEY SCHUTTE was sous-chef at the time at De Librije, a Michelin three-star restaurant housed in the library of a former fifteenth-century abbey in Zwolle, the Netherlands, and is now executive sous-chef at the Landmark Mandarin Oriental in Hong Kong.

PIA SNITKJÆR obtained her master's degree in food science and technology from the University of Copenhagen in 2005 and earned her doctorate in molecular gastronomy in 2010. The focus of her thesis was the study of the fundamentals of meat stock reduction using sensory and chemical analysis. Snitkjær cowrote a review paper on molecular gastronomy, which was published in *Chemical Reviews* in 2010. She is active in establishing teaching in gastronomy at the University of Copenhagen.

HERVÉ THIS founded the molecular gastronomy movement and molecular cuisine. He is a physical chemist at Institut National de la Recherché

Agronomique (INRA), France, and director of the Molecular Gastronomy Group, at the Laboratory of Chemistry, AgroParisTech. Also, This is a professor at AgroParisTech as well as scientific director of the Food Science & Culture Foundation (French Academy of Sciences). He has published several books and articles on molecular gastronomy, including *Molecular Gastronomy* (2006).

THOMAS M. TONGUE JR. is the director for product development at Innovating Food Processors, Inc., in Faribault, Minnesota. In his work of the past twenty-one years, he has focused on fluid-bed encapsulation and agglomeration technology, in both ingredient development and applications of encapsulated and agglomerated food, beverage, and pharmaceutical ingredients. Tongue's recent contributions to the industry include the development and commercialization of a number of functional powder beverages, agglomerated high-intensity sweeteners (rebaudioside A–based a component of stevia), and a range of encapsulated ingredients and new coating systems for the food, bakery, meat, snack foods, confection, and dietary-supplement industries.

JOB UBBINK is founder of Food Concept & Physical Design (www.themill.ch), a food-strategy and technology company based in Switzerland. Trained as a physical chemist at the University of Leiden, he obtained his doctorate at Delft University of Technology (both institutions are located in the Netherlands). He served as a visiting scientist at both Moscow State University and the University of Bristol, the United Kingdom. Ubbink worked for several years at the flavor company Givaudan in Switzerland and was for more than ten years at the Nestlé Research Center in Lausanne.

ERIK VAN DER LINDEN is a professor of food physics at Wageningen University, the Netherlands, where he initiated a molecular gastronomy program. Also, he directs a program at TIFN, the Dutch Top Institute for Food and Nutrition, on sensory and food structure. Previously, he worked for Unilever in the areas of detergents and cosmetics, both in the Netherlands and the United States. He has coauthored more than one hundred scientific papers.

PAULA VARELA currently works as researcher at the Institute of Agrochemistry and Food Technology, IATA-CSIC, Spain. She has been involved in a wide range of academic and industrial food-related projects within Europe and South America. Her area of expertise is sensory and consumer sciences and the physical properties of food. Lately, Varela's research has focused on the application and development of new methodologies to further understand consumer perception. She has published more than thirty scientific papers and contributed a range of book chapters.

CÉSAR VEGA is a senior applications scientist for Mars Botanical, a division of Mars Incorporated, in Rockville, Maryland, where he is in charge of designing novel food systems for the efficacious delivery of bioactive compounds. He earned his doctorate in food science from University College Cork, Ireland, in 2006. His area of expertise is in dairy science, particularly ice cream and spray-dried milk products, and more recently, the science of cooking. Vega is a passionate cook. He obtained culinary training from Le Cordon

Bleu Culinary Arts Institute in Ottawa, Ontario, and since then he has been extensively involved in all aspects related to the science of cooking. He served as one of the expert reviewers in Myhrvold's *Modernist Cuisine: The Art and Science of Cooking*.

THOMAS VILGIS, a physics professor at the University of Mainz, Germany, leads the Statistical Physics Group and the Experimental Soft Matter Food Science Group at the Max Planck Institute for Polymer Research. He has written and cowritten about 260 scientific papers and a book on the physics and materials science of nanocomposites. He has also published, in German, several books about molecular gastronomy and the scientific aspects of cooking and food, including three full-color cookbooks about the molecular kitchen. His book *Molekularküche* (2005) was granted the World Cookbook and Best Cookbook for Professionals award.

PETER WIERENGA is an assistant professor of food chemistry at Wageningen University, the Netherlands.

PETER J. WILDE is a senior scientist at the Institute of Food Research in the United Kingdom and research leader at the Food Structure and Health Program. He has a degree in biophysics from the University of East Anglia and obtained his doctorate in 2000 in the interfacial mechanisms underlying the stability of foams and emulsions.

CHRISTOPHER YOUNG is a chef–scientist known for applying science and technology to create culinary experiences that earlier generations could never have imagined. Before becoming a chef, Young completed degrees in mathematics and biochemistry. From 2003 to 2007, he worked with the world-famous chef Heston Blumenthal in overseeing culinary development at the Fat Duck restaurant in the United Kingdom. He wrote, with Maxime Billet and Nathan Myhrvold, *Modernist Cuisine* (2011), a groundbreaking book that explores how a deeper understanding of science and technology helps chefs achieve greater feats of creativity and innovation in their kitchens.

INDEX

Italic page numbers indicate material in tables or figures.

bacon, 73–82
bacteria: and cooking temperature for food safety, 170; and food preservation, 84–86
baguettes, 234
Baked Alaska recipe, 173
baker's yeast, 227
baking and meringue structure, 113–115
baking powder, 229
baking soda (sodium bicarbonate), 91–92, 96–97
baking stones, 232
banana, enzyme activity in, 48
Barbecue Chicken recipe, 172–173
batch freezers, 283
battered-and-breaded fried foods, 157, 158
battered food, cooking of, 160–161
Bavarian pretzels, 95–96
béchamel (white) sauce, 47, 49–50, 149
beef stock, 211
beer: foam, 102–103, 120; kriek, 130; as raising agent, 159
beta-myrcene, 36
betanine, 198, 199
biopolymers, 128
biting and perception of crispness, 162
bitterness, 94, 103, 196–204
blanching, 48
blind, eating, 233–236
bloomed chocolate, 65–72
Blumenthal, Heston, 17, 159, 277
boiling and phase change, 168–169
bones, stock from, 95, 206–212, 215
botulism, 92n1
bouillon, 206, 210
bread: basic ratio, 278; crumb coating, 159–160; crust, 95; flour, 226; staleness of, 55; thermal diffusivity of, 169–170
Brillat-Savarin, Jean-Anthelme, 1, 252
brine, 48, 75–76, 87
brown butter, 247, 280, 282–286

browning: bacon, 80; onions, 91–92, 96–97; pizza dough, 226. *See also* Maillard (browning) reaction
brown sauces, 95
bubbles: egg whites, 19; fermentation, 227–228; and foam, 118–119; formation of, 159; in ice cream, 127–130; sponge cake, 19; stability of, 101–106, 245; vapor, 168
butter: brown, 247, 280, 282–286; clarified, 98, 202, 284; in ice cream, 279–280, 282, 284–285; infused, 196–205; Kientzheim, 247; in sauce, 259; substituting olive oil for, 18
buttermilk, 234
butterscotch, 93–94
But the Crackling Is Superb (Kurti & Kurti), 4–5, 176
butyric acid, 282

cacao, 267, 286
cake flour, 226
calcium alginate network, 220
calcium bridges, in cheese, 7–9
calcium gluconolactate, 30
calcium sulfate, 221
California-style pizza, 225
Calippo Shots, 126–127
calorie-for-nutrient (CFN) index, 268
Camembert cheese, 198
candied bacon, 81
canola oil, 227
Cantonese cooking, 170–172
capillary effects, 247–248
cappuccino, 117, 120–121, 248
caramel candy, 93–94
caramelization reaction, 49, 56–57, 93, 189, 191, 286–287
carbohydrates: amorphous and glassy carbohydrates, 186–194; glucomannans, 35; in ice cream, 35–39, 41–43, 123–126, 128, 132; in milk, 7; storage, 209; and water, 158

carbon dioxide, 120, 159, 227–228

carboxymethylcellulose, 148

carotenoids, 57, 135

carrots, 11

casein, 7–9, 120

c* concentration, 149–150

cellulose, 77

Champagne, 95, 102

cheese: calcium bridges in, 7–9; fondue, 9; infused flavor in, 198; melting of, 8–9; mild vs. aged melt, 9

chemical gels, 249

chemical leavening, 228–229, 231

chemistry, food, 45–50, 169, 252

cherries, 130–131

Cherry-Kriek Sorbet recipe, 130–131

chewiness: dodol, 52, 55; konjac dondurma, 33–39, 41; pizza dough, 224–225

Chicago deep dish pizza, 225

chicken: barbecued, 172–173; breast, 178; consommé, 247; prefried nuggets, 162–163; roast, 167, 176; thighs, 164

Chinese cooking, 170–172

chocolate: bloomed, 65–72; conching, 267; forms of, 286–287; mousse, 281; pudding, 140; sugar in, 192–193

Chocolate Chip Cookies recipe, 63–64

clarified butter, 98, 202, 284

Clostridium botulinum, 92n1

coagulation, 158

coalescence, 101, 103, 105, 110, 119, 261

cocoa butter, 66, 71, 286

coconut milk, 57

coffee: aroma compounds, 199–200; beans, 94; extracts, 202–205

Coffee Butter recipe, 203

Coffee Ice Cream recipe, 204

Coffee Sauce recipe, 204

cohesiveness, 23

coldness, perception of, 126–127

collagen, 177, 209–210, 218, 247

collagenase, 212

colloids, 45, 49, 101

color: cues and perception, 16; darkening due to baking soda, 96; darkening from Maillard/caramelization reactions, 49, 57; darkening of cut foods, 48; extraction of betanine, 198; gelatin and, 259; vacuum-sealed dough, 62; when selecting bacon, 78–79

combination (combi) ovens, 184

conalbumin, 106

conceptual cuisine, 274–275

conching (in chocolate preparation), 267

condensed milk, 140, 286, 287

conduction, of heat, 167

conductivity rate (k), 168, 170, 172

connective tissue, 177, 209

consumer food science, 238–241

contrast, perceiving, 246

contrast theory, 238

convection, 167

convection ovens, 184

Cook & Chemist (Groenewold & Mariën), 196

cookies (biscuits): baking soda in, 96; blooming of chocolate chips in, 66–72; dough, 59–64; excessive browning in, 98; icing, 140

cooking: defined, 169, 251; fast, 169, 171; movements in, 274; from scratch, 264–272; slow, 169; *sous vide*, 95, 178, 184; as "spectator sport," 273–274. *See also* science and cooking

copper bowl (and beating of egg whites), 106, 270

Cordon Bleu School (Paris), 243

cornicabra olives, 19

cornichons, 147

cotton candy, 186–187, 188

crackling: defined, 13, 79; duck skin, 176–185; pork skin, 11, 13

cream, 285

creaminess, perception of, 261
creaming of bubbles, 104, 128
crème anglaise, 135, 280
crème brûlée, 135, 280
Crisp Battered Fish recipe, 16
crispness, 13, 16, 79–80, 155–157;
	analyzing, 162–163; baguette
	crust, 234; microwaving and, 161;
	perception of, 162, 260–261; puff-
	ing skin for, 179–185; water and,
	157–158
Crispy Battered Chicken Thighs
	recipe, 164
crispy–crunchy–crackly foods, 12–17,
	155–157
cross-linking (of biopolymers), 151,
	248
crunchiness: bacon, 74, 79–80; and
	crispness, 17, 79–80, 155–156;
	duck skin, 180; measuring sound
	of, 12–14; sodium acid pyrophos-
	phate, 229; vegetables, 171
crust: baguette, 234; crispy, 155–162;
	microwaveable browning, 96,
	164; pig trotters, 218–219; pizza,
	224–232
cryoconcentration, 213–214. *See also*
	freeze concentration
cryorendering, 181–183
cryosearing, 184–185
crystalline structure: cocoa butter
	polymorphs, 66; ice, 124–125;
	isomalt, 194; retrogradation, 55;
	sugar, 186, 188–189, 192–193
culinary pleasure, 254–263
cultural codes, 236
curd, 8–9
curing (of fish), 85–87
custard, 280

dark chocolate, 286
deaf people and food sounds, 15
decompression, 101
deep frying, 160, 162
defatted milk, 115

defrosting and food poisoning, 173
deglazing, 98
dehydration, 87, 97. *See also* drying;
	freeze drying
De Librije restaurant, 108, 115
demi-glaces, 95, 207–216
denaturing, 39, 48, 87, 121, 136
Denmark, 83–85, 87–89
density (ρ), 168
dielectric heating, 167n2
differential scanning calorimetry, 190
diffusion: gas, 28, 119; salts, 29–30,
	87; sugar aroma compounds, 192–
	193; of tastants to receptors, 150;
	water, 157–158
diffusivity, thermal (α), 168, 169–
	170, 174
disaccharides, 190, 194
disproportionation (of bubbles in
	foam), 119–120
dodol, 52–58
do-pyaza, 49
double extraction, 211
dough, cookie, 60–64
dry–crisp products, 157
dry-cured bacon, 74–76
dry ice, 181–183
drying, 85. *See also* dehydration
dry milk powder, 283
duck breast, 177–185
dulce de leche, 96, 285

effervescence, 102
efficiency in cooking, 60
efficiency of extraction, 201
eggbox model, for calcium binding by
	alginates, 27
eggplant, 46, 48
egg whites: bubbles in cake, 19; in
	copper bowl, 106, 270; and foam,
	249; and Maillard reaction, 93;
	meringue, 105, 110–111; whip-
	ping, 109–112, 121, 249, 270
egg yolks: emulsifying of, 111; fat in,
	282; flavor extraction with, 197;

infiltration of egg whites, 121; lecithin in, 159; texture, 134–141
elastic behavior, 50
elastic protein network, 118–121
El Bulli restaurant (Roses, Spain),
2–3, 26, 28, 277
emotions and senses, 256–257, 262
emulsifiers, 159, 193, 202, 248
emulsions: cream, 261, 285; egg
yolk, 135; oil-in-water, 42, 89,
101, 135, 245–246, 249; water-in-
water, 42
encapsulated bakery ingredients,
229–231
enfleurage, 197
entangling of polysaccharides, 150
enzymes, 212, 219–220
espresso, 203
essential oils, 57
evaporated/condensed milk, 285–287
evaporation, 164, 180, 184
expectations, effect on perceptions,
16, 237–241, 255–256
extensional viscosity, 153
extraction: comparing methods, 200,
203–204; solvent, 197–205; stock,
207–212
extraction efficiency, 201
extra virgin olive oil, 18–19

fast cooking, 169, 171
fat: in ice cream, 129; reducing in
food, 160; rendering, 178–183; in
stock extraction, 211–212
Fat Duck restaurant, 17, 159
fat migration, 67–72
fat-to-lean ratio (bacon), 78
fatty acids, 19n1
faux caviar, 30
fennel, 234–235
fermentation, 84, 102, 225–228
Fibrimex (FIB), 220–223
fibrinogen–thrombin networks, 220–
223
film deformation (in foams), 119

fish: and chips, 17; crisp battered, 16;
fried, 159; preservation techniques,
86–89; salmon, 84–85, 89, 234;
steamed, 194; sushi, 83–84, 89–90
flavor: balancing, 30; extraction,
202–205; groups, 200; release of,
65, 258–260; vs. taste, 247
flavor forms, 247
flour: in cake, 19; enriched, 265;
types of, 225–227
flow behavior: of ketchup, 145; of
molten sugar, 191; in oil–water
extraction, 201
foam, 101; batter, 159; béchamel,
50; beer head, 102–103, 120; eggs
and, 106, 249; film deformation,
119; formation of, 118–119; meringues as, 105–106, 109–114; microstructure, 118; milk, 117–122;
stability of, 119–120, 128; whipping, 50, 101, 104–105, 118–119,
248
foie gras Chantilly, 245
food, defined, 251
food-borne illness: and cooking temperature, 171; and preservation,
85–87
food poisoning, 173
Food Rules (Pollan), 264
food science, 265, 269–270, 276
"fox testicle" ice cream, 33
freeze-concentrated matrix, 124, 213
freeze concentration, 124. See also
cryoconcentration
freeze-dried reductions, 212–214
freeze drying, 212–215, 220
freezing point depression, 124–129,
131, 186, 282
French fries (chips), 11–12, 95
French Meringue recipe, 109
French-style ice creams, 33
frictional energy, 167n2
frozen desserts: "frozen Florida,"
173–174. See also ice cream
frozen meat and fish, 76, 84, 173

fructose, 56, 209
frying: bacon, 80–81; battered food, 160; duck, 180–185; fish, 159
furan, 88
furanone, 97
fusion cuisine, 26

Gagnaire, Pierre, 198, 243, 248, 277
galactomannans, 42–43, 148, 149
ganache, 285
gardening, 266–267
gases, 100–106
gas-phase extraction, 197
gastronomy, defined, 252
gelatin, 26; in ice cream, 128–132; jellified oil-in-water emulsions, 245–246; melted, 151, 219, 246; mouthfeel, 148–149; in pig's feet, 218–222; in sauce, 259; from stock extraction, 209–210; whipped foam, 248
gelatinization, 55, 160, 178–179, 183
gelato, 123
gelling agents, 25–27, 148
gelling bath, 31
gels, 35, 177
geolu, 135
ghee, 98
glaces, 207–216
glass transition temperature (T_g), 188–190, 193–194
glassy state: of bacon, 80; of meringue, 105, 113; of skin, 180; of sugar, 186–194; and water diffusion, 158
glazed meat, 93
glucomannans, 35–39, 41–42
glucose, 56, 281, 283
gluten, 225–227
glutinous rice flour, 54
glycerin, 248
glycogen, 209
Gordon-Taylor equation, 190
Gouda cheese, 9, 198
granddad sushi, 83

granita, 127
grapefruit wind crystals, 249
gravlax/gravadlax, 84–85
Gravlax recipe, 84
green ketchup, 146–147
green-olive spheres, 31
grilled cheese sandwich, 7–10
grilling, 80, 93, 218–219, 221–222
Gruyère cheese, 9
guar gum, 42, 148, 149
Guinness, 120
gula melaka, 56
gums, 25

hardness, 23
hard-wheat flour, 225–226
hardwood (and smoking), 76–77
haute cuisine, 274, 277
health issues, 264–272
heat capacity, 167–168
heat denaturation, 88
heat flux, 174
heating rate, 137
heat transfer, 166–174
hemicellulose, 77
herring, smoked, 83–88
high-gluten flour, 225
home economics, 271
Homemade Tomato Ketchup recipe, 143
honey, 26
host–guest relationship, 237
humidity, 184
hydrate shell, 186
hydration, 60–64
hydrocolloids, 35, 148
hydrolysis, 178, 210
hydrophilic molecules, 199
hydrophobic aroma compounds, 193
hydroxypropylmethylcellulose, 150

ice cream: butter in, 279–280, 282, 284–285; coffee, 196, 204; composition of, 34; formula/equation for, 124–126, 281–283; freezing point

depression, 124–126; as frozen foam, 34; ingredients of, 130–132; konjac dondurma, 33–40; melting behavior, 126–127; soda/float, 102; stretchy, 36–39; structure of, 110, 124, 129–130
ice spheres, 126–127
ice wine, 213
incompatible mixtures, 42–43, 71, 101
In Defense of Food (Pollan), 264
infantile amnesia, 257
infiltration, egg-yolk, 121
infusion, 196–205
ingredients: adapting to cooking method, 170–174; encapsulated, 229–231; pairing, 123–133; physical properties of, 169; ratios, 278, 282–283; selection for flavor balance, 30
injection: brine, 75–76; gas, 101
in-mouth sensorial perception, 260
internal setting gels, 220
Internet and cooking, 273, 281
inulin, 209
inverse spherification, 30
invert sugar, 281, 283
isomalt, 189, 194
isoviscosity, 140
Italian meringue, 109

jams, 98
Japanese cuisine, 41–42, 89
jellified oil-in-water emulsions, 245–246
Jell-O, 25–26
jelly agar, 57
jingaisitsu (Japanese teahouse), 255

kaiseki dinner, 256
ketchup, 142–147
Kientzheim butter, 247
kinesthesia, 259–260
Kitab al-Tabikh (al-Baghdadi), 46
konjac dondurma, 36–39

Konjac Dondurma recipe, 38
konjac glucomannan, 41–42
konnyaku, 36
kriek, 130
Kurti, Nicholas, 4, 174, 176, 243, 275

lactic acid, 8
lactose (milk sugar), 7–8, 194
latent heat, 168–169
Lavoisier, Antoine-Laurent de, 250
lax/laks, 84
leaching, 142–144
leavened yeast buns, 93
leavening, 96, 159, 227–231, 248
lecithin, 159
lignin, 77
lipophilic molecules, 199
liquefaction, sensation of, 187
liquid nitrogen, 37
liquid smoke extracts, 76
Little Miss Millet poem, 7, 9
lobster, 197, 201, 247–248
locust bean gum (LBG), 42–43, 148, 149
lox, 84
lye (sodium hydroxide, caustic soda), 95–96

Madeira cake, 175
Maillard, Louis-Camille, 91–92
Maillard (browning) reaction, 49, 56–57, 77; in brown butter, 284–285; in Champagne, 92; in chocolate, 286; pH dependency of, 91–92; reducing sugars and, 91, 93; in smoked fish, 88; speeding up, 92; during stock reduction, 215; temperature and, 94–95
Malaysia, 52, 56
Manchego cheese, 9
mange-tout (snap pea), 17
mango ravioli, 29
mannitol, 194
Maraş dondurma, 34–36

marinades: and baking soda, 92n1; barbecue, 93, 172; for herring, 83

marrow, 211

mass-produced foods, 267–269

mastication, 259–260

mastic resin, 36–38

matrix, food: cheese, 9; cocoa butter, 66; glassy sugar, 193; ice cream, 35, 43, 124, 128–129, 131; meringue, 105

mayonnaise, 111, 247

McGee, Harold, 4, 111, 135, 210, 277

MCP (monocalcium phosphate), 229

meat glues, 218–222

medallions, pig trotter, 218–223

Mediterranean sea bream, 245

Mediterranean sponge cake, 18–24

melanoidins, 92

melting: agar-agar gels, 26–27; cheese, 8–9; chocolate, 65–66, 69; fat, 178–183, 211, 229; gelatin, 151, 219, 246; ice cream, 123–126, 173–174; ice spheres, 126–127; sorbet, 131; sugars, 187–192, 194; temperature at, 169

menu-item description studies, 238–240

meringue: baked Alaska, 173–174; composition of, 105; physics and chemistry of, 110–114, 249

micelles, 8

microbial transglutaminase (TGase), 218–222

microorganisms: and cooking temperature, 172; and preservation, 85–89

Microwavable Meringues recipe, 105

microwave heating: baked Alaska, 174; battered-and-breaded foods, 160–163; and crispness, 163; how it works, 164–165; meringues, 105, 174; pies, 96

milk: composition of, 7, 49; converting into cheese, 8; evaporated/ condensed, 285–287; fat in, 71; foam, 117–122, 128; minerals in, 7; solids, 284; use in pastry, 279–287; whole, 120–121

milk-based meringue, 108

milk chocolate, 286

milk jelly, 248

minerals, 7, 209, 265

Minibar restaurant, 194

"miracle" foods, 265–266

miscibility, 191

mix-and-go dough, 229

Mixed-Vegetable Tartlet recipe, 146, 147

mixing behavior, 150–151, 191, 193

mojito, spherified, 28

molecular cuisine, 242–250

molecular gastronomy, 242, 270, 275, 277

molten sugars, 188, 191

monocalcium phosphate (MCP), 229

morello cherries, 130–131

mother sauces, 280

moussaka, 45–50

Moussaka recipe, 47–48

mousse, 281, 287

mouth coating, 149

mouthfeel. See texture (mouthfeel)

mucilage gums, 44

multisensorial integration, 258

Mustard Sauce recipe, 85

myoglobin, 78

Newtonian and non-Newtonian behavior, 148, 151

New York-style pizza, 225

nitrogen, 120, 127

nitrous oxide, 3, 20–21, 159

noisiness, 11–17

Noma restaurant (Copenhagen, Denmark), 239

Non-enzymatic browning. See Maillard (browning) reaction

nonessential nutrients, 265–266

nonfat dry matter, 212, 281–282

soups and sauces, 148–153
sour gherkins (cornichons), 147
sous vide cooking, 95, 178, 184
soy-based macaroon, 115
Space Dust candy, 14–15
Spain, 94, 217
sparging, 101
specific heat capacity (C_p), 167–168
speed ovens, 165
Spherical Mango Ravioli recipe, 29
spherification, 25–32
spider (ice cream float), 102
spinal tube defects (and vitamins), 265
sponge cake: black sesame, 2; foaming action, 105; Mediterranean, 18–24; olive oil, 20; syphoned, 20–24
sponginess (of cake), 19
springiness (of cake), 23
squid, 259
staleness (of bread), 55
starch: and crispness, 165; granules, 151, 153; thickening, 55, 149–152; and water, 158. *See also* amylopectin; amylose
steak, 94–95, 172, 285
steam: bubbles, 180; injection, 118, 120–121
steamed sea fish, 194
sterilization, 121
stifado, 49
Stir-Fried Sweet and Sour Pork recipe, 171
stir-frying, 170–172
stocks, 95, 206–216; amino acids and taste of, 247; extraction duration, 210–211; reduction, 212–214
strawberries: ice cream, 249; in ketchup, 147; in sorbet, 130
stretchy texture, 39, 41–44
stuffed pig trotters, 217–221
sublimation, 213–214
sucrose/saccharose (table sugar), 56, 189–194, 281

sugar-free ice cream, 125
sugars: blends, 189–193; crystalline and amorphous states, 188, 192–193; effect on water activity, 157; and freezing point depression, 124–125, 131; glassy state, 186–194; and meringue, 112–113; in pizza dough, 226–227; from stock extraction, 209; viscosity of, 128. *See also* carbohydrates
sunlight and oxidation, 85–87
superheated steam, 180–181
surface-active agents/surfactants, 103, 118, 120, 198, 202
surfaces: bubbles, 103–105; food, 95–96, 161
susceptor materials, 161–164
sushi, 83–84, 89
sweetened condensed milk, 140, 286, 287
sweetness, 189–190
Swiss meringue, 109
syphons, 159

table sugar (saccharose, sucrose), 56, 189–194, 281
tartaric acid, 9
tarts, 157
tastants, 142
taste panels, 13–14
taste release, 151
taste vs. flavor, 247
tea ceremony, 255–256
't Brouwerskolkje restaurant, 5, 115
technique, defined, 251–252
technology: defined, 252; ice cream and, 281
temperature: effect on flavor extraction, 201, 203–204; and foodborne illness, 170–171, 184n2; freezing, 126–128; and Maillard (browning) reaction, 94–95; thermal properties at room, 168
temperature-controlled water circulator/water bath, 136, 137

tempering, 66, 191
tempura, 158–159
terpenes, 36
texture (mouthfeel): boiled egg, 136–137; bread, 228; components of, 11; crispy and crunchy, 155–157; ice cream, 123–128; ketchup, 145; perception of, 258–262
texturizers, 131
thermal conductivity (k), 168, 170
thermal diffusivity (α), 168, 169–170, 174
thermal properties, 168
thickening: by alginic acid, 29; by egg yolk, 135, 137–141; by glucomannans, 37, 42; of ketchup, 142–147; by locust bean gum (LBG), 42–43, 148, 149; by salep flour, 34; of soups and sauces, 148; by starch, 55, 149–152; by whipping, 104; by xanthan gum, 31, 149
thick-sliced bacon, 80
thin-sliced bacon, 80
This, Hervé, 4, 111, 198, 275, 277
Three-Cheese Grilled Cheese Sandwich recipe, 9
toast, 157, 169–170
tocopherols and tocotrienols, 57–58
toffee, 93–94
tokonoma (alcove), 256
tomato ketchup, 142–147
tomato meringue, 115
Tom's Old World Pizza Crust with New World Ingredients recipe, 231
tongue, 198–199, 261
torrefacto/torrado, 94
torrefied coffee, 94
transfer rate, 201
transglutaminase (TGase), 218–222
trehalose, 189–194
trisodium citrate, 29
truffles, 197
Tselementes, Nikolaos, 46
tsukuda-sensei (master of the ceremony), 255–256

2-furaldehyde, 97
Two Cultures and the Scientific Revolution, The (Snow), 269
Tzatziki Ice Cream recipe, 132–133

ultrasound, 15
umami, 13
unsaturated fatty acids, 87

vacuum packaging, 74
vacuum sealers, 61–64
vanilla diplomat cream, 248
vanilla extraction, 198
vapor bubbles, 168
veal stock, 208
vegetable oil, 227
vegetables, raw
vicinal water, 180
virgin olive oil (VOO), 18–19
viscoelasticity, 35, 38
viscosity, 25–26; c* concentration, 149–150; in ice cream, 128; of ketchup, 143; of liquid foods, 148n1; of starch suspensions, 151; sugars and, 112, 186, 193; and taste perception, 152–153, 261; temperature and, 201; of various foods, 139
vitamins, 265
vocabulary of the kitchen, 263
vodka, 159
volatile aroma compounds, 187, 199, 207
volume heating, 167
von Liebig, Justus, 208, 245

water: component of skin, 177; diffusion of, 157–158; evaporation, 164, 180, 184, 248; freezing point of, 124–126, 282; in pizza dough, 226; steam, 159; vicinal, 180
water activity (a_w), 157–158
water-and-oil phase extraction, 198–205
water bath, 136, 137

water-in-water emulsions, 42
waxy starch, 54, 151–152
well-tempered chocolate, 66
wet–crisp products, 157
wet-curing bacon, 73, 75–76
wheat starch, 150–151
whey, 7–8, 120
whipped-cream: Chantilly, 245; dairy-based, 104; dispenser/syphon, 3, 20–24; Pavlova, 104
whipping: egg whites, 109–112, 121, 249, 270; foam, 50, 101, 104–105, 118–119, 248; making butter, 285
white bacon, 74
white beet mayonnaise, 198
white chocolate, 286
whole milk, 120–121

wicking, 179
Wiltshire curing, 75
wind crystals, 249
wine, 9, 213
Wöhler sauce, 243, 247
wok cooking, 170–172
women, role of, 270–271
wood, 76–77, 94

xanthan gum: and ketchup, 142–147; in sponge cake, 20–24; and viscosity, 30, 149
Xanthomonas campestris, 144

yeast, pizza, 225, 227
yeast buns, 93
yield stress, 146

Arts and Traditions of the Table: Perspectives on Culinary History
Albert Sonnenfeld, Series Editor

Printed in the USA
CPSIA information can be obtained
at www.ICGtesting.com
LVHW041738021223
765352LV00001B/20

9 780231 153454